5-15-18

I0396736

# Human Errors

# Human Errors

*A Panorama of Our Glitches,*
*From Pointless Bones*
*to Broken Genes*

## Nathan H. Lents

Houghton Mifflin Harcourt
Boston   New York
2018

*Library of Congress Cataloging-in-Publication Data*
Names: Lents, Nathan H., author.
Title: Human errors : a panorama of our glitches, from pointless bones to
broken genes / Nathan H. Lents.
Description: Boston : Houghton Mifflin Harcourt, 2018. |
Includes bibliographical references and index.
Identifiers: LCCN 2017046396 (print) | LCCN 2017057190 (ebook) |
ISBN 9781328974693 (hardback) | ISBN 9781328974679 (ebook)
Subjects: LCSH: Human physiology. | Human evolution. | BISAC: SCIENCE /
Life Sciences / Human Anatomy & Physiology. | SCIENCE / Life Sciences /
Evolution. | SCIENCE / Life Sciences / Biology / General.
Classification: LCC QP34.5 .L467 2018 (print) | LCC QP34.5 (ebook) | DDC
612—dc23
LC record available at https://lccn.loc.gov/2017046396

Book design by Chloe Foster
Illustrations by Donald Ganley/© Houghton Mifflin Harcourt
Charts © Houghton Mifflin Harcourt

Printed in the United States of America
DOC 10 9 8 7 6 5 4 3 2 1

Now, there's a topic you know a lot about!

—My mother upon learning I was writing a book about human flaws

# Contents

refrain, however imperfectly. The song itself was composed by another such human (albeit an exceptional one), a person who I dare to suggest had little appreciation for the genes, proteins, and neurons that were hard at work as he did so.

Despite the fact that we often take them for granted, the capabilities of the human body are simply wondrous—miraculous even. So why not write a book about *that?*

Because you've heard about it many times. Those books have already been written. If you want a book about the glorious intricacy of the human body, you are in luck—simply walk into any medical library, and you will find tens of thousands of volumes. If you count biomedical journals, where new discoveries are announced, the number of accolades to the greatness of the human form rises into the tens of millions. There is no shortage of words and pages dedicated to how well the body usually works.

This is not that story. This is a story of our many flaws, from head to toe.

As it turns out, our flaws are extremely interesting and informative. By exploring human shortcomings, we can peer into our past. Each and every flaw discussed in this book tells a story about our species' evolutionary history. Every cell, every protein, and every letter in our DNA code has been subjected to the harshness of natural selection over the fullness of evolutionary time. All of that time and all of that selection has resulted in a body form that is fantastically robust, strong, resilient, clever, and mostly successful in the great rat race of life. But it is not perfect.

We have retinas that face backward, the stump of a tail, and way too many bones in our wrists. We must find vitamins and nutrients in our diets that other animals simply make for themselves. We are poorly equipped to survive in the climates in which we now live. We have nerves that take bizarre paths, muscles that attach to nothing, and

# Introduction:
# Behold the Blunders of Nature

Here is a story you've heard numerous times: Behold the incredible beauty, complexity, and greatness of the human body and its many systems, organs, and tissues! It seems that the deeper into our bodies we look, the more beauty we find. Like the layers of an onion, the cells and molecules that make up the human body have seemingly infinite levels of complexity. Human beings enjoy a rich world of the mind, perform astoundingly complex physical tasks, digest food material and then commingle it with their own matter and energy, effortlessly turn genes on and off, and every now and then produce whole new individuals in "endless forms most beautiful."

Somehow, all of these processes come together to create the wondrous complexity of human life while allowing us to remain oblivious to the underlying mechanisms. A perfectly ordinary human can sit down and play "Piano Man" without ever having to think about the cells and the muscles of her hands, the nerves in her arms, or the brain centers where the information to play the piece are stored. A second human can sit and listen to the song without ever bothering to contemplate the vibrations of his eardrum, the conductance of nerve impulses to his auditory processing center, or the memory recall that allows him to belt out the lymph nodes that do more harm than good. Our genomes are filled with genes that don't work, chromosomes that break, and viral carcasses from past infections. We have brains that play tricks on us, cognitive biases and prejudices, and a tendency to kill one another in large numbers. Millions of us can't even reproduce successfully without a whole lot of help from modern science.

Our flaws illuminate not only our evolutionary past but also our present and future. Everyone knows that it is impossible to understand current events in a specific country without understanding the history of that country and how the modern state came to be. The same is true for our bodies, our genes, and our minds. In order to fully grasp any aspect of the human experience, we must understand how it took shape. To appreciate why we are the way that we are, we must first appreciate what we once were. To twist the old saying a bit, we can't understand where we are now if we don't know where we came from.

Most of the human design flaws that I describe in this book fall into one of three categories. First, there are aspects of our design that evolved in a different world than the one we now inhabit. Evolution is messy and takes time. Our species' tendency to gain weight easily and lose weight only with difficulty made very good sense in the Pleistocene savannas of Central Africa but not so much in a twenty-first-century developed nation.

The next category of flaws includes those of incomplete adaptation. For example, the human knee is the product of a redesign that took place as our ancestors gradually shifted from a quadrupedal posture and an arboreal lifestyle to a bipedal posture and a mostly terrestrial lifestyle. Most of the various components of the knee adapted very well to the new demands placed on this crucial joint, but not all of the kinks were worked out. We are almost fully adapted to upright walking—but not quite.

The third category features those human defects that are due to nothing more than the limits of evolution. All species are stuck with the bodies that they have and they can advance only through the tiniest changes, which occur randomly and rarely. We inherited structures that are horrendously inefficient but impossible to change. This is why our throats convey both food and air through the same tiny space and why our ankles have seven pointless bones sloshing around. Fixing either of those poor designs would require much more than one-at-a-time mutations could ever accomplish.

A good example of the tremendous constraints of evolution even during episodes of great innovation is the vertebrate wing. Wings have been invented in many separate lineages. The wings of bats, birds, and pterosaurs all evolved separately and therefore have big structural differences. However, in all of those cases, the wing evolved from a forelimb. Those animals lost many uses of their forelimbs in order to get wings. Neither birds nor bats can grasp things very well. They have to crudely use their feet and mouths to manipulate objects. It would have been far better for those animals to grow wholly new wings while retaining their forelimbs, but evolution rarely works that way. For an animal with a complex body plan, growing new limbs is not an option, but slowly reshaping existing limbs is. Evolution is a constant game of tradeoffs. Most innovations come with a cost.

Evolutionary innovations are as varied as they are pricey. They range from copying errors in the blueprints inside each cell to glaring design defects in the assembly of bones, tissues, and organs. In this book, I will address each of these categories of errors in turn, looking at whole sets of flaws that share the same general themes and that, taken together, tell an incredible story about how evolution works, what happens when it doesn't, and the high price our species has paid for these adaptations over the millennia.

Human anatomy is a clumsy hodgepodge of adaptations and maladaptations. We have pointless bones and muscles, underwhelming senses, and joints that don't quite keep us upright. Then there's our diet. Whereas most animals do just fine eating the same thing day in and day out, we humans have to have ridiculously varied meals in order to get all the nutrients we need. Most of the contents of our genomes are completely useless, and occasionally they are actually harmful. (We even carry around thousands of dead viruses tucked in the DNA of every one of our cells, and we spend our lifetimes dutifully replicating these carcasses.) And there are yet other, even more astounding imperfections: We are incredibly inefficient at our ultimate goal of making more of ourselves, and we have immune systems that attack our own bodies, just one of many design-related diseases. Even what is arguably our crowning evolutionary achievement—the powerful human brain—is filled with defects that lead individuals to make extremely poor choices in their daily lives, sometimes at the expense of their very existence.

But as strange as it may sound, there is beauty in our imperfections. How boring would our lives be if each of us were a purely rational, perfect specimen? Our flaws are what make us who we are. Our individuality comes from tiny variations in our genetic and epigenetic codes, and much of this diversity arises from the haphazard insults of mutation. Mutations, like lightning strikes, are random and often destructive, but they are also, somehow, the source of all human greatness. The flaws discussed in this book are scars of battles won in the great struggle for survival. We are the unlikely survivors of this endless evolutionary conflict, the products of four billion years of dogged perseverance in the face of great odds. The history of our flaws is a war story unto itself. Gather round and listen.

# Human Errors

# 1

## Pointless Bones and Other Anatomical Errors

*Why the human retina is installed backward; why one of our mucous drains is located near the top of a sinus cavity; why our knees are so bad; why the disks of cartilage between our vertebrae sometimes "slip"; and more*

We love to admire physical excellence. We can't get enough of massive bodybuilders, graceful ballerinas, Olympic sprinters, shapely swimsuit models, and hardy decathletes. In addition to its innate beauty, the human body is also dynamic and resilient. The carefully orchestrated functions of the heart, lungs, glands, and GI tract are truly impressive, and we continue to discover the elaborate intricacies through which the body maintains its health despite the onslaught of a changing environment. Any discussion of the shortcomings of our physical form must first begin with an acknowledgment that the beauty and capability of the human body far outshines the few odd quirks here and there.

But quirks there definitely are. Lurking in our anatomy are some odd arrangements, inefficient designs, and even outright defects. Mostly, these are fairly neutral; they don't hinder our ability to live and thrive. If they did, evolution would have handled them by now. But some are not neutral, and each has an interesting tale to tell.

Over millions of generations, human bodies morphed tremendously. Most of our species' various anatomical structures were transformed in that metamorphosis, but a few were left behind and exist now purely as anachronisms, the whispers of days long gone. For instance, the human arm and the bird wing perform totally different functions but have striking structural similarities in the scaffolding of their bones. That's no coincidence. All quadruped vertebrates have the same basic skeletal chassis, modified as much as possible for each animal's unique lifestyle and habitat.

Through the random acts of mutation and the pruning of natural selection, the human body has taken shape, but it's not a perfect process. A close inspection of our mostly beautiful and impressive bodies reveals mistakes that got caught in one of evolution's blind spots—sometimes literally.

## I Can't See Clearly Now

The human eye is a good example of how evolution can produce a clunky design that nonetheless results in a well-performing anatomical product. The human eye is indeed a marvel, but if it had been designed from scratch, it's hard to imagine it would look anything like it does now. Inside the human eye is the long legacy of how light-sensing slowly and incrementally developed in the animal lineage.

Before we consider the puzzling *physical* design of the eye, let me make one thing clear: The human eye is fraught with *functional* problems as well. For instance, many of the people who are reading this book right now are doing so only with the aid of modern technology. In the United States and Europe, 30 to 40 percent of the population have myopia (nearsightedness) and require assistance from glasses or contact lenses. Without them, their eyes do not focus light properly, and they cannot make out objects that are more than a few feet away. The rate of myopia increases to more than 70 percent of the population in Asian countries. Nearsightedness is not caused by injury. It's a design defect; the eyeball is simply too long. Images focus sharply before they reach the back of the eye and then fall out of focus again by the time they finally land on the retina.

Humans can also be farsighted. There are two separate conditions that cause this, each resulting from a different design flaw. In one, hyperopia, the eyeballs are too short, and the light fails to focus before hitting the retina. This is the anatomical opposite of myopia. The second condition, presbyopia, is age-related farsightedness caused by the progressive loss of flexibility of the lens of the eye, the failure of the muscles to pull on the lens and focus light properly, or both. Presbyopia, which literally translates as "old-man sight," begins to set in around age forty. By the age of sixty, virtually everyone has difficulty making out close objects. I'm thirty-nine, and I have noticed that I hold books and newspapers farther and farther from my face each year. The time for bifocals is nigh.

Add to these common eye issues others such as glaucoma, cataracts, and retinal detachment (just to name a few), and a pattern begins to emerge. Our species is supposed to be the most highly evolved on the planet, but our eyes are rather lacking. The vast majority of people will suffer significant loss of visual function in their lifetimes, and for many of them, it starts even before puberty.

I got glasses after my first eye exam, when I was in the second grade. Who knows how long I had actually needed them? My vision isn't just a little blurry. It's terrible — somewhere around 20/400. Had I been born before, say, the 1600s, I would probably have gone through life unable to do anything that required me to see farther than arm's length. In prehistory, I would have been worthless as a hunter — or a gatherer, for that matter. It's unclear if and how poor vision affected the reproductive success of our forebears, but the rampant nature of poor vision in modern humans argues that excellent vision was not strictly required to succeed at least in the most recent past. There must have been ways that early humans with poor vision could have thrived.

Human vision is even more pitiable when compared with the excellent vision of most birds, especially birds of prey such as eagles and condors. Their visual acuity at great distances puts even the sharpest human eyes to shame. Many birds can also see a broader range of wavelengths than we can, including ultraviolet light. In fact, migrating birds detect the North and South Poles *with their eyes*. Some birds literally *see* the Earth's magnetic field. Many birds also have an additional translucent eyelid that allows them to look directly into the sun at length without damaging their retinas. Any human attempting to do the same would most likely suffer permanent blindness.

And that's just human vision during the day. Human night vision is, at best, only so-so, and for some of us it is very poor. Compare ours with cats', whose night vision is legendary. So sensitive are cats' eyes that they can detect a single photon of light in a completely dark environment. (For reference, in a small, brightly lit room, there are about one hundred billion photons bouncing around at any given moment.) While some photoreceptors in human retinal cells are apparently able to respond to single photons, these receptors cannot overcome background signaling in the eye, which leaves humans functionally incapable of sensing just one photon and thus unable to perform the sorts of visual feats that cats pull off so easily. For a human to achieve conscious perception of the faintest possible flash of light, she needs five or ten photons delivered in rapid succession, so cats' vision is substantially better than humans' in dim conditions. Furthermore, human visual acuity and image resolution in dim light is far worse than that of cats, dogs, birds, and many other animals. You might be able to see more colors than dogs can, but they can see at night more clearly than you.

Speaking of color vision, not all humans have that either. Somewhere around 6 percent of males have some form of colorblindness. (It's not nearly as common in females because the screwed-up genes that lead to colorblindness are almost always recessive and on the X chromosome. Because females have two X chromosomes, they have a backup if they inherit one bum copy.) Around seven billion people live on this planet, so that means that at least a quarter of a billion humans cannot appreciate the same palette of colors that the rest of the species can. That's almost the population of the United States.

These are just the *functional* problems with the human eye. Its physical design is riddled with all sorts of defects as well. Some of these contribute to the eye's functional problems, while others are benign, if befuddling.

One of the most famous examples of quirky design in nature is the retina of all vertebrates, from fish to mammals. The photoreceptor cells of vertebrate retinas appear to be installed backward — the wiring faces the light, while the photoreceptor faces inward, away from it. A photoreceptor cell looks something like a microphone; the "hot" end has the sound receiver, and the other end terminates in the cable that carries the signal to the amplifier. The human retina, located in the back of the eyeball, is designed such that all of the little "microphones" are facing the wrong way. The side with the cable faces forward, toward the light, while the hot end faces a blank wall of tissue.

This is not an optimal design for obvious reasons. The photons of light must travel around the bulk of the photoreceptor cell in order to hit the receiver tucked in the back. When you're speaking into the wrong end of a microphone, you can still make it work, provided that you turn the sensitivity of the microphone way up and you speak loudly, and the same principles apply for vision.

Furthermore, light must travel through a thin film of tissue and blood vessels before reaching the photoreceptors, adding another layer of needless complexity to this already needlessly complicated system. To date, there are no workable hypotheses that explain why the vertebrate retina is wired backward. It seems to have been a random development that stuck because correcting it would be very difficult to pull off with sporadic mutations—the only tool evolution has in its toolkit.

This reminds me of the time when I installed a piece of molding called a chair rail—molding that's placed about halfway up a wall—in my house. It was my first attempt at woodworking, and it didn't go as well as I had hoped. The long pieces of wood for a chair rail are not symmetrical; you have to choose which is the top surface and which is the bottom, and unlike with crown molding or baseboards, with chair rails, it's not immediately obvious which is the top and which is the bottom. So I just picked the way I thought looked best and then set about installing it: doing all the measurements, making the cuts, staining the wood, hanging it, sinking the nails, applying wood putty to the seams and nail holes, and staining again. Finally, I was done. The first guest to see my handiwork immediately pointed out that I had installed the chair rail upside down. There *was* a correct top and bottom and I got it wrong.

This is a good analogy to the backward installation of the retina because way back in the beginning, the light-sensing patch of tissue that would evolve into the retina could have faced in either direction with little functional difference for the organism. As the eye continued to evolve, however, the light sensors moved inside the cavity that would become the eyeball, and the backward nature of the installation became clear. But it was too late. At that point, what could be done? Flipping the whole structure around could never be achieved through a couple of mutations here and there, just like I could not simply flip my chair rail around; all the cuts and seams would be inverted. There was no way to correct my blunder without starting over completely, and there is no way to correct the backward installation of the vertebrate retina without starting over completely. So I kept the upside-down chair rail, and our ancestors kept their backward retinas.

Interestingly, the retina of cephalopods—octopi and squid—is not

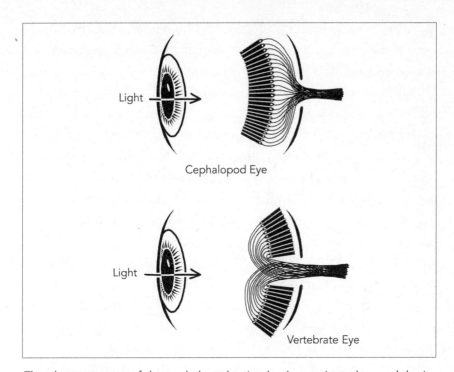

Cephalopod Eye

Vertebrate Eye

The photoreceptors of the cephalopod retina (top) are oriented toward the incoming light, while those of the vertebrate retina (bottom) are not. By the time this suboptimal design became disadvantageous to vertebrates, evolution was powerless to correct it.

inverted. The cephalopod eye and the vertebrate eye, while strikingly similar, evolved independently of each other. Nature "invented" the camera-like eye at least twice, once in vertebrates and once in cephalopods. (Insects, arachnids, and crustaceans have an entirely different type of eye.) During the evolution of the cephalopod eye, the retina took shape in a more logical way, with the photoreceptors facing toward the light. Vertebrates were not so lucky, and we are still suffering from the consequences of this evolutionary fluke; most ophthalmologists agree that the backward retina is what makes retinal detachment more common in vertebrates than in cephalopods.

There is one more design quirk in the human eye that merits mention. Right smack in the middle of the retina, there is a structure called the optic disk; this is where the axons of the millions of photoreceptor cells all converge to form the optic nerve. Imagine the tiny cables from millions of tiny microphones all coming together into one bundle, a cable of cables to carry all the signals to the brain. (The visual center of the brain happens to be in the very back, as far from the eyes as it could possibly be!) The optic disk, located on the surface of the retina, occupies a small circle in which there are no photoreceptor cells. This creates a blind spot in each eye. We don't notice these blind spots because having two eyes compensates for it; our brains fill in the pictures for us, but they are definitely there. You can find simple demonstrations of this on the Internet by searching for *optic disk blind spot.*

The optic disk is a necessary structure insofar as the retinal axons must all converge at some point. A much better design would be to place it deeper in the back of the eye, tucked underneath the retina rather than smack on top of it. However, the backward placement of the retina makes the blind spot unavoidable, and all vertebrates have it. Cephalopods do not, because the right-side-in retina allows the easy placement of the disk behind an unbroken retina.

Perhaps it would be too greedy to ask for hawks' eyes, but couldn't we aspire to those of an octopus at least?

## Nasal Sinuses That Drain *Upward*

Just below the eyes, you'll find another set of evolutionary errors: the nasal sinuses, a meandering collection of air- and fluid-filled cavities, some of them deep inside our heads.

Many people don't appreciate just how much open space there is in the skull. When you inhale through your narrow nostrils, the flow of air branches into four pairs of large chambers tucked in the bones of your face; this is where the air comes in contact with mucous membranes. The mucous membranes are highly folded patches of wet and sticky tissue designed to catch dust and other particles, including bacteria and viruses, so that they do not reach your lungs. In addition to trapping particulates, the sinuses are also useful for warming and humidifying the air you breathe.

The mucous membranes in the nasal sinuses produce a slow and steady flow of sticky mucus. This mucus is swept away by tiny, pulsating, hairlike structures called cilia. (Picture a miniature version of the hairs on your arms constantly swirling in order to push sticky water along your skin.) Inside your head, mucus drains into several spots and is ultimately swallowed and sent to the stomach — the safest place to put the mucus, since the bacteria and viruses it contains can be dissolved and digested by the acid there. The sinus passages, when working properly, keep the mucus flowing, which clears the bacteria and viruses before they can cause infections and prevents mucus from gumming up the whole system.

Of course, the whole system does get gummed up sometimes, and that can lead to a sinus infection. Bacteria that are not swept along fast enough can set up camp and establish an infectious colony that may spread throughout the sinuses and beyond. Mucus, normally thin and mostly clear, becomes thick, viscous, and dark green when you have an infection. Most infections are not serious, but they aren't fun either.

Have you ever noticed that dogs, cats, and other animals don't seem to have head colds nearly as often as humans do? Most humans suffer between two and five head colds (also called upper respiratory infections) per year, and some of them are accompanied by full-blown sinus infections. In the six years that I have had my dog, however, I haven't noticed a single episode of a runny or stuffy nose, watery eyes, coughing, or repeated sneezing. He's never even had a fever that I know of. Sure, dogs *can* get sinus infections, and the most common symptom is the easily recognized runny nose. But this is a rare occurrence for them. Most dogs will go their whole lives with no major episodes of infection in their nasal sinuses.*

Wild animals are similarly free of nasal symptoms. Sinus infections are possible but rare in nonhuman animals, although they are a little more common in primates than in other mammals. Why do humans have it so bad?

There are a variety of reasons for why we're so susceptible to sinus infections, but one of them is that the mucous drainage system is not particularly well designed. Specifically, one of the important drainage-collection pipes is installed near the *top* of the largest pair of cavities, the maxillary sinuses, located underneath the upper cheeks. Putting the drainage-collection point high within these sinuses is not a good idea because of this pesky thing called gravity. While the sinuses behind the forehead and around the eyes can drain downward, the largest and lowest two cavities must drain upward. Sure, there are cilia to help propel the mucus up, but wouldn't it be easier to have the drainage below the sinuses rather than above them? What kind of plumber would put a drainpipe anywhere but at the bottom of a chamber?

This poor plumbing is not without consequence. When the mucus becomes thicker, things get sticky, both figuratively and literally. Mucus thickens when it carries a heavy load of dust, pollen, or other particulates or antigens; when the air is cold or dry; or when a bacterial infection is fighting to take hold. During these times, the cilia have much more work to do to get the sludgy mucus to the collection point.

---

* We must except from this statement dog breeds with truncated snouts, such as Pekingese and pugs. These are the product of intense artificial selection by human breeding, not natural selection. In fact, most health problems suffered by dogs in general are the result of recent selective inbreeding and not common in their wolf ancestors.

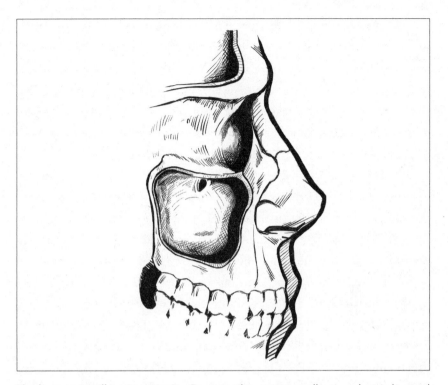

The human maxillary sinus cavity. Because the mucous collection duct is located at the top of the chamber, gravity cannot help with drainage. This is part of the reason why colds and sinus infections are so common in humans but unheard of in other animals.

If only we had gravity to help with the drainage, like other animals do! Instead, our cilia must work against gravity *and* the increased viscosity of the thick mucus. They simply can't keep up and this leads to the nasal symptoms of the common cold. This is also why colds and allergies occasionally trigger secondary bacterial sinus infections; as the mucus pools, bacteria can fester.

The poor location of the drainpipes in the maxillary sinuses also helps to explain why some people with colds and sinus infections can briefly find relief by lying down. When they don't have to work against gravity, the cilia in the maxillary sinuses can propel some of the thick mucus toward the collecting duct, which relieves some of the pressure. This is no cure, however, and the respite is only temporary. Once a bacterial infection takes hold, drainage alone can no longer combat it, and the bacteria must be defeated by the immune system. In some people, mucous drainage is so poor that only nasal surgery can bring relief from near-constant sinus infections.

But *why* is the drainage system at the top of the maxillary sinuses instead of below? The evolutionary history of the human face holds the answer. As primates evolved from earlier mammals, the nasal features underwent a radical change in structure and function. In many mammals, smell is the single most important sense, and the structure of the entire snout was designed to optimize this sense. This is why most mammals have elongated snouts: to accommodate huge air-filled cavities chock-full of odor receptors. As our primate ancestors evolved, however, there was less reliance on smell and more reliance on vision, touch, and cognitive abilities. Accordingly, the snout regressed, and the nasal cavities got smushed into a more compact face.

The evolutionary rearrangement of the face continued as apes evolved from monkeys. The Asian apes—gibbons and orangutans— simply ditched the upper set of cavities altogether; their lower sinuses are smaller and drain in the direction of gravity. The African apes— chimpanzees, gorillas, and humans—all share the same type of sinuses. However, in the other apes, the sinuses are larger and more cavernous, and they are joined to each other by wide openings, which facilitates unrestricted flow of both air and mucus. Not so with humans.

Nowhere are there more differences between humans and nonhuman primates than in the facial bones and skull. Humans have much smaller brows, smaller dental ridges, and flatter, more compact faces. In addition, our sinus cavities are smaller and disconnected from one another, and the drainage ducts are much skinnier. Evolutionarily speaking, humans gained nothing by having those drainage pathways squeezed into narrow tubes. This was likely a side effect of making room for our big brains.

This rearrangement produced a suboptimal design that has left us more susceptible to colds and painful sinus infections than perhaps any other animal. But as far as poor design goes, this evolutionary mishap is nothing compared to what lurks just a bit farther down in the body: a nerve that should drive straight from the brain to the neck but instead takes a few dangerous detours along the way.

## A Runaway Nerve

The human nervous system is astonishingly intricate and important. Our brains are highly developed, and our nerves make those brains functional.

Nerves are bundles of tiny individually wrapped cables called axons that convey impulses from the brain to the body (or, for sensory nerves, from the body to the brain). For example, there are motor neurons that live near the top of the brain that send their long axons out of the brain, down the spinal cord, out of the lumbar region, and down the legs to their targets in the big toes. A long route, for sure, but a direct one. There is a web of cranial and spinal nerves that carry their axons from the brain to every muscle, gland, and organ in the body.

Evolution has left us with some very bizarre defects in that system. Consider just one example, the awkwardly named recurrent laryngeal nerve (RLN). (There are actually a pair of these nerves, one on the left and one on the right, as there are for most nerves in the human body, but for simplicity's sake, we'll talk about just the left one.)

The axons found in the RLN originate near the top of the brain and connect to the muscles of the larynx (also known as the voice box). These muscles, under the direction of the nerve, are what allow us to make and control audible sounds when we speak, hum, and sing.

You would think that axons that begin in the brain and end in the

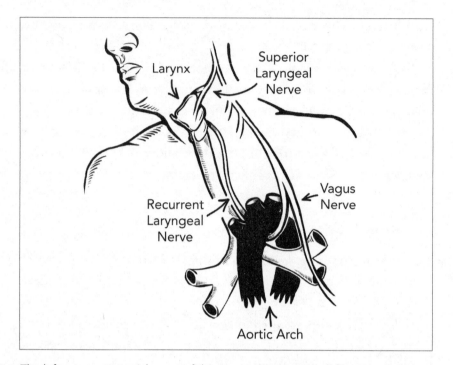

The left vagus nerve and some of the nerves that branch off from it—including the recurrent laryngeal nerve (RLN). Its circuitous route through the chest and neck is an evolutionary throwback to our early vertebrate ancestors, in which a straight path from the brain to the gills went very near to the heart.

upper throat would travel a short distance: through the spinal cord, into the throat, and to the larynx. The whole thing could be just a few centimeters long.

Nope. The axons of the RLN are packaged within a more famous nerve, the vagus. It travels down the spinal cord all the way to the upper chest. From there, the sub-bundle of axons known as the RLN exit the spinal cord a little below the shoulder blade. The left RLN then loops under the aorta and travels back up to the neck, where it reaches the larynx.

The RLN is more than three times longer than it has to be. It winds through muscles and tissue that it need not. It is one of the nerves that heart surgeons must be very careful with, given how it intertwines with the great vessels from the heart.

This anatomical oddity has been recognized as far back as the time of the ancient Greek physician Galen. Is there a functional reason for this circuitous route? Almost certainly not. In fact, there is another nerve, the superior laryngeal nerve, that also innervates the larynx and travels the exact route that we would predict. This sub-bundle, which also branches from the larger vagus bundle, leaves the spinal cord just underneath the brain stem and travels the short distance to the larynx. Nice and easy.

So why does the RLN travel this long, lonely road? Once again, the answer is in ancient evolutionary history. This nerve originated in ancient fish, and all modern vertebrates have it. In fish, the nerve connects the brain to the gills, which were the ancestors of the larynx. However, fish don't really have necks, their brains are tiny, they don't have lungs, and their hearts are more like muscular hoses than pumps like ours. Thus, a fish's central circulatory system, located mostly in the space behind its gills, is quite different than a human's.

In fish, the nerve makes the short trip from the spinal cord to the gills in a predictable and efficient route. Along the way, however, it weaves through some of the major vessels that exit the fish heart, the equivalent of the branching aortas of mammals. This weaving makes sense in fish anatomy and allows for a compact and simple arrangement of nerves and vessels in a very tight space. But it also paved the way for an anatomical absurdity that would begin to develop as fish evolved into the tetrapods that would eventually give rise to humans.

During the course of vertebrate evolution, the heart began to move farther back as the body form took on a distinct chest and neck. From fish to amphibians to reptiles to mammals, the heart inched farther and farther away from the brain. But the gills did not. The anatomical position of the human larynx relative to the brain is not that different from the relative position of fish gills to the brain. The RLN should not

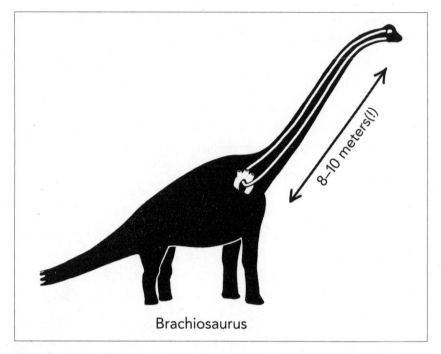

Brachiosaurus

The left recurrent laryngeal nerve (RLN) loops under the aorta in all vertebrates. Therefore, the RLN of sauropod dinosaurs would have been incredibly long.

have been affected by the changing position of the heart — except that it was intertwined with the vessels. The RLN got stuck and was forced to grow into a large loop structure in order to travel from the brain to the neck. Apparently, there was no easy way for evolution to reprogram the embryonic development of this nerve so as to untangle it from the aorta.

The result of this is that the RLN forms a long, unnecessary loop in the human neck and upper chest. While this may not seem like a huge deal, consider that all of the tetrapod vertebrates are stuck with the same anatomical arrangement inherited from a common ancestor, bony fishes. The RLN of the ostrich should need to travel only two to three centimeters to do its job, but instead, it travels nearly a full meter down the spinal cord, then a full meter back up the neck. The RLN of the giraffe can be up to five meters long! Of course, that is nothing compared with how long the RLN must have been in the apatosaurus, brachiosaurus, and other sauropods. Maybe we shouldn't scorn our own relatively puny RLN after all.

## Pains in the Neck

A runaway nerve is just one thing that is messed up with the human neck. Really, the whole neck is a bit of a disaster. For starters, it's very poorly protected, especially compared to the protection other important areas get. Just above the neck, the brain is kept in a thick, rigid housing that can withstand a substantial degree of trauma. Below the neck, the heart and lungs are protected by a strong but flexible rib cage anchored by a flat chest plate that is similarly sturdy. Evolution went to a lot of trouble to protect the brain and the cardiopulmonary system but left the connections between them totally vulnerable. (It also failed to protect our visceral organs well, but that is a story for another day.)

It is very difficult for someone to do great harm to your brain or heart with his bare hands, but your neck can be snapped with one swift motion. This weakness is not unique to humans, but humans do have special problems. For example, the vertebrae that are so good at allowing smooth movement as we twist and turn our necks are also easily dislocated. The trachea—the tube through which fresh air gets to the lungs—rests just underneath a thin layer of skin in the front of the neck and can be pierced by even a dull point with little force. The human neck is just a glaring vulnerability.

An even more basic flaw in the neck is the fact that there is one tube, from the opening of the mouth until about halfway down the neck, that is common to both the digestive and respiratory systems. The throat conveys both food and air—what could possibly go wrong? While this, too, isn't an exclusively human problem—the throat is a nearly universal structure in birds, mammals, and reptiles—that doesn't make it less of a flaw. In fact, this universal poor design demonstrates the physical constraints that evolution has to work with. Mutations are good for making small incremental tweaks but they cannot be used to execute full-scale redesign. Most of the higher animals get their food and air through the same tube. Having totally separate anatomical structures for digestion and respiration would make much more sense in terms of hygiene, immune defense, and the general maintenance of these very different systems—but evolution came up with a different, less sensible solution for many animals, humans included.

For breathing, especially, our bodies are supremely underequipped. Air passes down a single tube in the throat that then divides into dozens of branches in the lungs. These branches terminate in tiny dead ends full of air sacs that allow the exchange of gases across a thin membrane. The pathway for expired air is precisely the reverse. Air comes in and out like ocean tides through all those branches, hence humans are termed *tidal breathers.* This is horribly inefficient because there is a great deal of stale air left in the lungs when fresh air is brought in. These mix, diluting the oxygen content of the air that actually reaches the lungs. This burden of stale air in the lungs limits oxygen delivery, and so we must overcome this by breathing deeper, particularly during moments of peak demand, like exercise.

To get an exaggerated sense of the extra work humans must do because of tidal breathing, try breathing through a tube or a hose. But don't try it for too long, because if the tube is more than a few feet, you will slowly suffocate, no matter how deeply you breathe. If you have ever snorkeled, you have experienced the same exaggerated effects of tidal breathing. Even when floating quietly, making only gentle motions with their legs and arms, snorkelers must breathe deeply to remain comfortable. Every breath is a mixture of stale and fresh air. The longer the path, the more stale air left behind at the end of every breath.

There is a much better way to breathe. In many birds, the airway splits into two lanes of traffic before it reaches the respiratory sacs. Inbound air heads directly to the lungs, not mixing with stale air. Stale air is collected into an out spout and travels upward, joining with the trachea only high in the throat. One-way flow of traffic into the lungs ensures that each breath brings in mostly fresh air. This is a much more efficient design and allows birds to take far shallower breaths than we do to deliver the same amount of fresh air into their bloodstream. This is a critical improvement for birds, because flying demands massive quantities of oxygen.

Of course, the biggest danger in the human throat's design is not suffocating, but choking. Nearly five thousand Americans choked to death in 2014, the majority of them choking on food. If we had separate openings for air and food, this would never happen. Cetaceans —whales and dolphins—have blowholes, a powerful innovation that provides a dedicated conduit for air. Many birds and reptiles also have a superior design for breathing in that their nostrils convey air directly to the lungs rather than merging with the throat. This is why snakes and some birds can continue to breathe even while they are slowly working on swallowing a huge meal. Humans and other mammals have no such apparatus; when we are swallowing, we have to stop breathing momentarily.

It also doesn't help that the human instinctual physical reaction when startled is to gasp. That in and of itself is an example of poor design. What is the benefit of suddenly and forcefully pulling in a large breath of air when you are frightened or receive surprising news? There's no upside to that, and if there is food or liquid in your mouth at that moment, it can create a big problem.

Even though all mammals can get foreign bodies lodged in their tracheae, humans are particularly prone to choking because of some very recent evolutionary changes in our species' neck anatomy. In other apes, the larynx is substantially lower in the neck than ours. This design allows for a longer throat, giving more room for the muscles involved in swallowing to do their work. In all mammals, during swallowing, a flap of cartilage (called the epiglottis) must slap down over the opening of the trachea to cover it so food heads to the stomach, not the lungs. Of course, this usually works just fine, but not always, and recently the human voice box has been drifting upward, shortening the throat and tightening the space in which the delicate dance of swallowing is done.

Most scientists believe that the larynx has migrated high in the neck of the modern human to enhance vocalizations. With shallower throats, humans are able to contort their soft palates in ways that other apes cannot, giving us a far richer toolkit for making sounds. Indeed, many of the vowel sounds found throughout world languages today are made possible only by our species' unique throat. There is even one specific sound, the throat click (made by the tight puckering of the back of the throat), that only humans can make and that is a standard part of many sub-Saharan African languages. While it's a bit too strong to say that our throats evolved purely or mostly to enable this click sound, it was one of a variety of vocalizations that were made possible by the gradual elevation of the voice box.

But these unique vocal powers came at a cost. The rise in the larynx meant the squishing of the throat, causing swallowing to be a much more error-prone affair. For babies, swallowing can be truly hazardous because there is just not much room in their tiny throats to accomplish the complicated and coordinated muscle contractions that this basic act involves. Anyone who has cared for infants or toddlers knows that they choke on their food and drink constantly, but that doesn't happen much in other young animals.

Swallowing is a good example of the limits of Darwinian evolution. The human throat is simply too complex for a random mutation — the basic mechanism of evolution — to undo its fundamental defects. We have to resign ourselves to the absurdity of taking in air and food through the same pipe.

A different evolutionary dynamic helps to explain the next design flaw, which also concerns one of the most elementary human activities: moving around on two feet. Here, the issue isn't that evolution *can't* solve the problem but that it simply hasn't—at least, not yet. The problem is due to incomplete adaptation. Nowhere is this clearer than in the human knee.

## Knuckle-Walkers

Whereas other primates move about on all four limbs, humans walk on two legs; this is called bipedalism. If you watch gorillas, chimps, and orangutans when they're not swinging from trees, they amble about using their feet and their knuckles. Sure, they can stand up on two legs and clumsily walk that way for short stretches, but it is not comfortable for them, and they're not good at it. Human anatomy, however, has evolved to support our species' standing upright, mostly by way of changes in the legs, pelvis, and vertebral column. We move much faster this way, and it's inefficient to move around on four limbs. We must have perfected the bipedal posture by now, right?

Not so much. The anatomical adaptation to upright walking never quite finished in humans. We have several defects that are the result of this failure to complete the process. For example, the intestines and other visceral organs are held together with thin sheets of connective tissue called mesenteries. Mesenteries are elastic and act to keep the gut loosely in place. However, these thin sheets are not suspended from the top of the abdominal cavity, as would make sense for a bipedal posture. Instead, they are attached to the back of the abdominal cavity, like they are in the other apes. That makes good sense for our quadrupedal cousins, but it's a poor design for us and causes occasional problems.

People who sit upright with little movement for long periods strain these mesenteries, which can then tear, requiring surgery. This defect hasn't yet been corrected by evolution because the selective pressure to

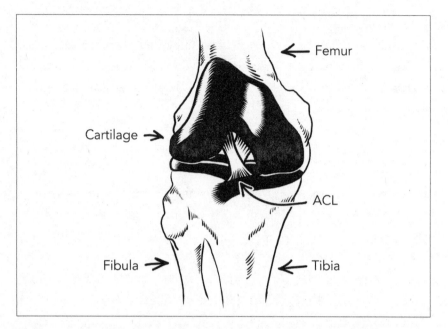

The bones and ligaments of the human knee, with the kneecap (patella) re-moved to reveal the anterior cruciate ligament (ACL). Our incomplete adap-tation to bipedalism has forced this relatively skinny ligament to endure much more strain than it is designed for, which is why humans—athletes especially —suffer torn ACLs so often.

fix it is quite low; before truck driving and desk jobs rolled around, torn mesenteries were probably quite rare. Still, this is poor design, leading to an unnecessary convolution of connective tissue in our abdomens.

There are more serious examples as well. Have you heard of the an-terior cruciate ligament? If you are a sports fan, you have; the tearing of this ligament (referred to as the ACL) is one of the most frequent sports injuries. Probably most common in football, torn ACLs also occur in baseball, soccer, basketball, track and field, gymnastics, tennis—basi-cally all high-impact, fast-paced sports. Located in the middle of the knee, the ACL connects the femur (thighbone) to the tibia (shinbone) and is located underneath the patella (kneecap), deep inside the joint. It does most of the work of holding the upper leg and lower leg together.

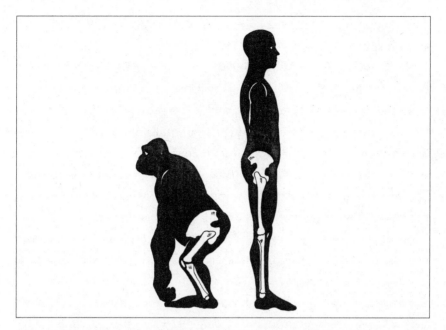

The natural postures of a standing ape and a standing human. Because of our erect bipedal posture, humans rely on our leg bones to bear most of our weight when standing and walking. Apes, on the other hand, often employ a bent-leg posture, which recruits muscles to share the burden.

The ACL is vulnerable to tearing in humans because our upright, bipedal posture forces it to endure much more strain than it is designed to. In quadrupeds, the strain of running and jumping is spread among four limbs, and the limb *muscles* absorb most of it. Once our ancestors transitioned to bipedalism, however, the strain was spread over two legs instead of four. This was too much for the muscles by themselves, so our bodies recruited the leg bones to help with the strain. The result was that human legs became straightened so that the bones, rather than the muscles, could bear most of the impact. Compare a standing human with a standing ape: a human's legs are fairly straight, while an ape's legs are bowlegged and usually bent.

This straight-leg arrangement works out okay for normal walking and running. But for sudden shifts in direction or momentum—when you're running and then stop short or when you make a sharp turn at high speed—the knees must bear the force of this sudden, intense strain. Sometimes, the ACL is simply not strong enough to hold the leg bones together as they twist or pull away from each other, and it tears.

To make matters worse, we as a species are getting heavier, so it's even harder for the ACL to withstand the strain put on it during those sudden shifts. This is especially true for athletes, who now weigh more than ever before and who also make lots of sudden high-speed weight shifts. You may have noticed that ACL injuries have become more common in the world of professional sports as athletes get ever larger.

Short of losing weight, we can't do much about this problem. It is not possible to isolate the ACL and strengthen it with exercise. It is what it is. Repeated strain doesn't make it stronger; it makes it weaker. As if that weren't bad enough, when the ACL is torn, it must be repaired surgically. Knee surgery demands a long recovery and rehabilitation period because ligaments are not very vascular—that is, they are fed by very few blood vessels and have very few of the cells that normally do the work of healing and rebuilding tissues. This is why ACL tears are among the most feared injuries in professional sports. A torn ACL usually means a full season lost.

The Achilles tendon holds another story about our imperfect evolution. No other nonskeletal structure underwent as dramatic a change during our species' transition to upright walking than this very conspicuous tendon. As our ancestors gradually shifted their weight from the balls of their feet to their heels, the Achilles tendon—which connects the calf muscle to the heel of the foot—found itself with much more work to do. A dynamic sinew, it responded well and is now the most visible feature of the human ankle. It has expanded dramatically to take on its demanding new role, and it reacts to both endurance exercise and strength training by becoming stronger still. The Achilles tendon is a workhorse.

However, by taking on most of the strain of the ankle joint, the

Achilles *tendon* has become the Achilles' *heel* of the entire joint, if you'll pardon the cliché. Injuries to the Achilles tendon are another one of the most frequent sports injuries, and there is no built-in redundancy for this tendon as there is in other joints. To add to the problems, the tendon is exposed prominently on the back of the leg, unprotected.

If the tendon is injured, even walking is impossible. The poorness of this design can be summed up in the observation that the function of the entire joint rests on the actions of its most vulnerable part. A modern mechanical engineer would never design a joint with such an obvious liability.

The knees and ankles aren't the only structures that underwent re-designs as our ancestors started to walk upright. The back also had to adjust. Ironically, as posture straightened, the back had to become *curvier,* particularly the lower back, which took on a pretty sharp concave shape in order to help transmit upper-body weight to the pelvis and legs evenly. Evolution even added bones to the lower back to allow for the sharper curve. Because of this curve, however, the lower back has to flex when you stand erect for long periods and it can fatigue. Lower-back pain is a common complaint among those whose jobs require them to stand in one place for hours.

Lower-back fatigue is mild compared to other problems we can have with our backs, and some of them are caused directly by design flaws. All vertebrates have disks of cartilage that lubricate the joints between the vertebrae in the spinal column. These disks are solid but compressible to absorb shock and strain. They have the consistency of firm rubber and allow the spine to be flexible while remaining strong. In humans, though, these disks can "slip" because they are not inserted in a way that makes sense given our species' upright posture.

In all vertebrates except us, the spinal disks are positioned in line with the normal posture of that animal. For example, the spinal columns of fish endure completely different kinds of strain than the spinal columns of mammals. The fish uses its backbone to stiffen its body and then

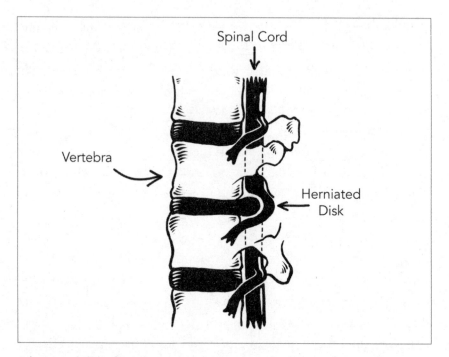

A herniated disk of cartilage in the human spinal column. As our ancestors adopted a more upright posture, the lumbar area of the vertebral column became sharply curved. The disks of cartilage between each vertebra are not optimally placed for this upright, curved posture; as a result, they sometimes "slip," leading to this painful condition.

pulls against it in a side-to-side motion in order to swim. But fish don't have to worry much about gravity and shock absorption since they are suspended in water. Mammals, however, must use limbs to hold their body weight, and those limbs must attach to the spinal column. Different mammals have different postures and so require different strategies for weight distribution via the spine. In almost all of the tremendously diverse spinal columns found in nature, the spinal disks have adapted to the posture and gait of the animal. But not in ours.

In humans, the vertebral disks are in an arrangement that is optimal for knuckle-draggers, not upright walkers. They still do a decent job of lubricating and supporting the spine, but they are much more prone to being pushed out of position than the vertebral disks of other animals. They are structured to resist gravity by pulling the vertebral joints toward the chest, as if humans were on all fours. With our upright posture, however, gravity often pulls them backward or downward, not toward the chest. Over time, this uneven pressure creates protuberances in the cartilage. This is known as a spinal disk herniation or, more commonly, a "slipped disk." Spinal disk herniation is nearly unheard of in any primate species but us.

Our ancestors began walking upright about six million years ago. It was one of the first physical changes as they diverged from other apes. It's disappointing, although not altogether surprising, that human anatomy has not had time to catch up and complete this adaptation. However, at least we use all the bones that we have in our backs. As mentioned earlier, as humans evolved to stand upright, a couple of bones were added to the lower back. Apparently, evolution can duplicate bones when needed. It seems it's not as good at deleting them when they're no longer needed.

## No Bones About It

Humans have way too many bones. This flaw isn't unique to us. Nature is replete with animals that have bones they don't need, joints that don't flex, structures that aren't attached to anything, and appendages that cause more problems than they're worth. The reason for this is that embryonic development is extremely complicated. For a body to take shape, thousands of genes must be activated and deactivated in a precise order, perfectly coordinated in time and space. When a bone, for example, is no longer needed, deleting it is not as easy as flipping a switch. Hundreds — maybe thousands — of switches must be flipped, *and* they must be flipped in such a way as to not screw up the thousands of other structures that are also built with those same genes. Remember, too, that natural selection flips these switches randomly, like a chimpanzee at a typewriter. If we wait long enough, the chimp will write a sonnet, but the wait will be long indeed. For anatomy, the result is a whole lot of baggage lying around.

For humans, some of the most striking anatomical redundancies are found in our skeletons. Take the wrist. It's a capable joint, no doubt about it. It can twist nearly 180 degrees in all rotational planes despite the vessels, nerves, and other cords that run from the arm to precise places in the hand. However, it is way more complicated than it needs to be. There are *eight* different bones in the wrist, not including the two bones of the forearm and the five bones of the hand. The small area that is just the wrist itself has *eight* fully formed and distinct bones tucked in there like a pile of rocks—which is about how useful they are to anyone.

Collectively, these wrist bones are helpful, but they don't really do anything individually. They sort of sit there when you move your hand. Yes, they connect the arm bones to the hand bones through a complex system of ligaments and tendons, but this arrangement is incredibly complicated and redundant. Redundancy *can* be a desirable thing, as we saw with our poor Achilles tendon, but not in the case of bones. Having extra bones requires many more attachment points for tendons, ligaments, and muscles. Each one of those points of contact is a weakness, a potential for strain or (as happens with the ACL) a debilitating tear.

We have examples of superbly designed joints in our bodies; the shoulder and hip joints come to mind. Not the wrist, though. No sane engineer would design a joint with so many individual moving parts. It clutters up the space and restricts the range of motion. If the wrist were rationally designed, it would allow the hand a full range of motion so that the fingers could bend backward and lie along the top of the arm. But of course it can't do that. The flexibility of the wrist joint is *restricted* by the many bones in there, not *facilitated* by them.

The human ankle suffers from the same clutter of bones that we find

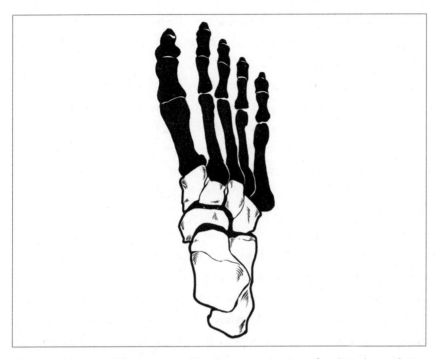

The seven bones of the human ankle (shown in white) are fixed in place relative to each other. No engineer would design a joint with so many separate parts, only to fix them together—yet incredibly, most humans manage just fine with this jumbled arrangement.

in the wrist. The ankle contains seven bones, most of them pointless. The ankle certainly has more to deal with than the wrist does, given that it is constantly bearing weight and is central to the locomotion of the entire body. But this is all the more reason why we would be better served by a simpler joint. Because many of the bones of the ankle do not move relative to one another, they would function better as a single, fused structure, their ligaments replaced with solid bone. Thus simplified, the ankle would be much stronger, and many of their current points of potential strain would be eliminated. There is a reason that twisted and sprained ankles are so common: the skeletal design of the ankle is a hodgepodge of parts that can do nothing except malfunction.

While the wrist bones and ankle bones are the most obnoxious examples of bones for which we have no use, there are others. For example, the tailbone.

The coccyx is the terminal part of the spinal column and consists of the last three (or four or five, depending on which you count) vertebrae fused together in a C-shaped structure. This section of bone has no function in humans. It doesn't house or protect anything; the spinal cord, which vertebrae are designed to protect, terminates much higher than where the coccyx begins. It is vestigial—a remnant from our ancestors who had tails.

Nearly all vertebrates have tails, including most primates. The great apes are among the rare exceptions, but even apes *begin* their embryonic lives with prominent tails. That tail eventually shrinks, and by the twenty-first or twenty-second gestational week, its vestige has become the worthless coccyx. Attached to the coccyx there is even the tiny remnant of a muscle—the dorsal sacrococcygeal muscle—that could flex the tailbone if the bones weren't fused. A pointless muscle for this pointless cluster of bones.

The coccyx does retain some connections to nearby musculature. It also bears much weight while you are in a reclined or seated position. But for the rare people whose tailbones are surgically removed because of injury or cancer, there are no long-term complications.

The human skull, like that of other vertebrates, is also a strange mishmash of bones that fuse together during childhood to form a single structure. The average human skull has twenty-two bones (some people have more!), with a lot of duplication. That is, the skull often has right-side and left-side versions of bones—a right and left jawbone fused together in the middle, for example, and a right and left top palate. There's no clear reason for this redundancy. While it makes sense that arms are separate structures, the same can't be said of the bones underneath the upper-lip area.

As with the duplicated bones in our skulls, there is no real reason to have paired bones in our forearms and lower legs. The upper arm has one bone, but the lower arm has two. Same for the leg; the thigh has one bone, but the shin has two. Yes, the two bones in the lower arm do allow for a twisting motion, but that's not the case in the lower leg. You cannot twist your leg below the knee without breaking something. Even in the forearm, having two parallel bones is not the only way to make a joint that can twist. In fact, having two bones ensures that the twisting cannot possibly exceed 180 degrees since the bones unavoidably knock into each other when you twist them. For comparison, the shoulders and hips do the task of twisting even better than the elbow, and they do it without the two-bone arrangement. No robot arm will ever be designed to imitate our nonsensical bone structure.

Human anatomy is beautiful, no doubt about it. We are very well adapted to our environment, but we are not *perfectly* adapted. Little imperfections exist. It's possible that, if our ancestors had lived the hunter-gatherer life for a longer time before moving into the modern era of vaccines and surgery, evolution would have continued to perfect human anatomy. However, that environment, like all environments, was so dynamic that evolution would simply have substituted our current imperfections for others. Evolution is a continual process — never quite complete. Evolution and adaptation are more like running on a treadmill than running on a track: we *must* keep adapting in order to avoid extinction, but it can feel like we never really get anywhere.

## Coda: A Dolphin with Hind Fins

Although we humans have our share of superfluous bones, there are many other animals with even more blatant examples of vestigial structures and extra bones. For example, there are some species of snakes that have tiny vestiges of a pelvis, though their limbs were lost eons ago. These useless snake pelvises don't attach to anything and perform no function. Then again, they don't really harm the animal; if they did,

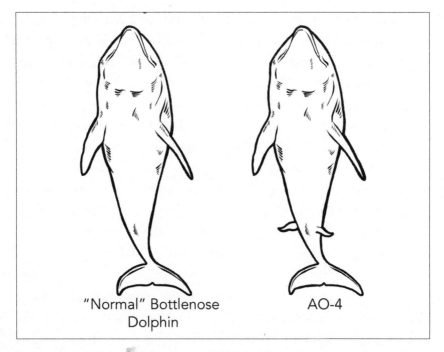

"Normal" Bottlenose Dolphin          AO-4

The "hind fins" of the dolphin named AO-4 (right), compared to a typical dolphin (left). These tiny but otherwise well-formed fins likely represent a spontaneous mutation that undid a prior one that caused the disappearance of the hind fins. Such "spontaneous revertants" offer a rare glimpse into how adaptations emerge through random mutations.

natural selection would have completed their removal from the body plan. Most whales also have the internal remnants of a pelvis — the quiet whispers of their legged ancestors that wandered back into the ocean more than forty million years ago. When these ancestors returned to marine life, their forelimbs gradually evolved into pectoral fins. Their hind limbs, however, simply regressed into nothing.

In 2006, Japanese fishermen caught a dolphin that had tiny hind fins, for lack of a better term. This dolphin, later named AO-4, was a rare find and was sent to the Taiji Whale Museum for display and further study.

The discovery of a dolphin with a tiny but perfectly formed pair of hind fins reveals the power of single mutation during development. In this case, a random mutation just happened to undo a previous mutation. Obviously, these are rare events—rather like lightning hitting the same place twice—but they are powerfully informative when we do find them. As of this writing, there has been no reported discovery of the precise mutation responsible for AO-4, but the scientific hunt continues.

It appears that the hind fins of dolphins didn't slowly regress in tiny increments down to nothing. Rather, a single mutation was able to take the last dramatic step and cause them to disappear entirely. Similar kinds of "high-impact" mutations were almost certainly responsible for the duplication of vertebrae in our species' lower back when we needed more of them for upright posture. Don't believe me? Humans are born every day with extra fingers or toes, perfectly formed and functional. If having twelve fingers had conferred a big advantage some time in our evolutionary past, you can bet that everyone would have twelve fingers now. Genes important for embryonic development have far-reaching effects and so mutations in just the right spot can make large anatomical rearrangements. These rearrangements are random and therefore usually result in harmful birth defects, but when we're talking about evolutionary timescales, events that seem unimaginably rare are possible.

Mutations like AO-4's lift the evolutionary veil that normally conceals an animal's past life. The mutation-driven tweaks and tugs of evolution can sometimes be undone, with dramatic results. Because we are constantly reminded of the slow and steady pace of evolution, we don't normally think of it as dramatic. The dolphin AO-4 reminds us that, at times, it can be.

# 2

# Our Needy Diet

*Why humans, unlike other animals, require vitamins C and $B_{12}$ in their diets; why almost half of all children and pregnant women are anemic despite getting plenty of iron; why we are all doomed to calcium deficiency; and more*

A casual stroll through a bookstore or library will reveal shelves upon shelves of books about food and eating. There are books on the history of cooking, books on exotic and ancient foods, cookbooks —and, of course, advice guides and manuals for fad diets.

We are constantly reminded of all the various things we need to eat. You must eat enough vegetables. Don't forget the fruits. A balanced breakfast is important. Remember to get lots of fiber. Meat and nuts are important for protein. Be sure to get omega-3 fatty acids. Dairy is important for calcium. Leafy greens are vital for magnesium and B vitamins. You cannot stay healthy by eating the same thing all the time. You should maintain a diverse diet in order to get all the various nutrients that your body needs.

And then there are the supplements. Now, most scientists consider the supplement industry a sham (I'm looking at you, herbal supplements), but many of these pills and powders do contain essential vitamins and minerals of which we simply must consume a minimum amount in order to be healthy. Some people's diets don't give them everything they need, and even people who get everything they need can't always absorb it properly. So sometimes, we need a little boost. That's why we're always being told to drink milk, for instance; it gives us the calcium that we need but can't produce in sufficient quantities ourselves.

Now compare our demanding diet with the diet of the cows that produce that milk. Cows can survive on pretty much nothing but grass. They live long and perfectly healthy lives and produce delicious milk and rich meat. How can these cows thrive without a delicate mix of legumes, fruits, fiber, meat, and dairy like humans are told to eat?

Forget cows; look at your own cats or dogs. Consider how simple their diets are. Most dog food is nothing more than meat and rice. No vegetables. No fruits. No supplemental vitamins. Dogs do just fine on this diet and, if not overfed, can live long and healthy lives.

How do these animals do it? Simple: they are better designed for eating.

Humans have more dietary requirements than almost any other animal in the world. Our bodies fail to make many of the things that other animals' do. Since we don't make certain necessary nutrients, we have to consume them in our diet or we die. This chapter tells the story of all the things we need to have in our diets simply because our lackluster bodies can't make them for us, substances as basic as, say, vitamins.

## The Scurvy of Humanity

Vitamins are what is known as *essential micronutrients,* a category of molecules and ions that we must get from our diets or we will suffer and die. (Other essential micronutrients are minerals, fatty acids, and amino acids.) Vitamins are among the largest molecules that cells need to survive.

## MAJOR DIETARY VITAMINS
## AND THEIR DEFICIENCIES

| VITAMIN | ALIAS | DEFICIENCY |
| --- | --- | --- |
| A | Retinol | Vitamin A deficiency |
| B$_1$ | Thiamine | Beriberi |
| B$_2$ | Riboflavin | Ariboflavinosis |
| B$_3$ | Niacin | Pellagra |
| C | Ascorbic acid | Scurvy |
| D | Cholecalciferol | Rickets, Osteoporosis |

Major dietary vitamins and the conditions that result from deficiency. Because humans adapted to thrive on a highly varied diet, we now need a highly varied diet in order to obtain all the micronutrients that we no longer synthesize in sufficient quantities for ourselves.

Most vitamins assist other molecules to facilitate key chemical reactions inside our bodies. For example, vitamin C assists at least eight enzymes, including three that are necessary for the synthesis of collagen. Even though we have these enzymes, they cannot make collagen without vitamin C. When the enzymes can't work, we get sick.

Vitamin C is called *essential* not because it is important but because we must get it from our diets. All vitamins are important, crucial even, to human health, but those that are essential are the ones that we cannot make ourselves and therefore must ingest.

In addition to vitamin C, there are other essential vitamins that perform important functions in the body. The B vitamins, for example, aid in the extraction of energy from food. Vitamin D helps us absorb and use calcium. Vitamin A is crucial for the functioning of the retina, and vitamin E has a wide variety of roles throughout the body, including protecting tissues from free radicals, harmful byproducts of chemical reactions.

The one thing that this diverse family of molecules has in common is that our bodies cannot make them. This is what makes vitamins A, B, C, D, and E different than, say, vitamin K or vitamin Q. If you haven't heard of those vitamins, it's because they are not *essential*, in the dietary sense of the word. They are just as important as other vitamins, but we don't need to get them from our food because we make them ourselves.

When people can't produce a specific vitamin and can't get it through their food, their health can really, *really* suffer. Again, vitamin C offers a useful example.

Schoolchildren in the United States often begin their study of American history by learning about the fifteenth- and sixteenth-century Europeans who explored this continent. I distinctly remember a story my classmates and I were told about how sailors carried potatoes or limes on their long voyages in order to prevent scurvy. As we now know, this horrible disease is caused by a deficiency in vitamin C. Without it, we cannot make collagen, an essential component of something called the extracellular matrix, or ECM. The ECM is like a microskeleton that runs through all of our organs and tissues, giving them shape and structure. Without vitamin C, the ECM gets weak, tissues lose integrity, bones become brittle, we bleed from various orifices, and our bodies basically fall apart. Scurvy is a dystopian novel written by the human body.

So how can dogs live on meat and rice, neither of which has any vitamin C whatsoever, and not develop scurvy? They make their own. In fact, nearly all animals on the planet make plenty of their own vitamin C, usually in their livers, and thus have no need for it in their diets. Humans and other primates are nearly alone in the need for dietary vitamin C (although guinea pigs and fruit bats have this problem as

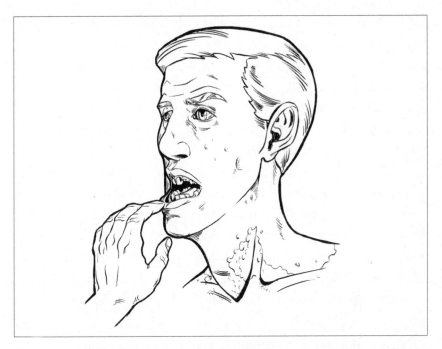

The physical appearance of scurvy. This horrific disease is caused by a deficiency of vitamin C—an essential micronutrient that human ancestors were able to make for themselves but that humans now must obtain through diet.

well). This is because, somewhere in our evolutionary past, human livers actually *lost* the ability to make this micronutrient.

How did we lose the ability to make vitamin C? Well, it turns out that we do have all of the genes that are necessary for vitamin C synthesis, but one of them is broken, mutated to the point of being nonfunctional. The broken gene, known as *GULO,* codes for an enzyme that is responsible for a key step in the manufacture of vitamin C. Somewhere in the ancestors of primates, the *GULO* gene suffered a mutation, rendering it inoperable, and then random mutation continued, littering the gene with tiny errors. As if to mock the uselessness of pieces of DNA like this, scientists call them *pseudogenes.*

We can still easily recognize the *GULO* gene in the human genome.

It's there, and the vast majority of the code is the same as in other animals, but there are a few key parts that have been mutated. It's as if you removed the spark plug from a car. It's still a car. You can easily see that it is still a car. In fact, you would have to look very carefully to find anything wrong with it at all. But it cannot function as a car, not even slightly. It's totally inoperable, even though most of it is still exactly the way it was before it was broken.

That's what happened to the *GULO* gene, way back in prehistory. The spark plug was removed by a random mutation. Over the course of evolutionary time, random mutations like this occur constantly. Often, they are of no consequence, but sometimes they strike right in a gene. When that happens, it is almost always bad, because the mutation usually disrupts the functioning of the gene. In such cases, the individuals in whose genomes these mutations occurred are a little worse off — or a lot worse off, if the change brings on a deadly genetic condition like sickle cell anemia or cystic fibrosis.

Often, the deadliest of these mutations are eliminated from the population when the people who carry them die. This begs the question: Why wasn't the *GULO* gene mutation eliminated? Scurvy is fatal. The consequences of this mutation ought to have been quick and harsh and should have prevented the harmful error from spreading throughout the species.

Well, maybe not. What if this disrupting mutation happened in a primate who, purely by chance, already had lots of vitamin C in her diet? For her, there would be no consequence of losing the ability to make vitamin C since she already ate foods that contained it. (What foods contain a lot of vitamin C? Citrus fruits. And where do citrus fruits mostly grow? Tropical rainforests. And where do most primates live? Bingo.)

The reason that the ancestors of primates could tolerate a mutation in the *GULO* gene was that, with plenty of vitamin C in their diets anyway, scurvy wasn't an issue. Since that time, primates — with the exception of humans—have pretty much stuck to rainforest climates. This preferred habitat is both a cause and a consequence of their inability to make vitamin C. After all, while it's easy to *break* a gene by mutation, it's much more difficult to *fix* it. It's like slamming the computer when it's not working right. Sure, you *might* fix it, but more likely, you'll harm it.

Primates aren't totally unique for having a screwed-up *GULO* gene. A few other animals have one as well. Not surprisingly, the ones that tolerate having the broken gene are the ones that get plenty of vitamin C in their diets. Take fruit bats, for example. They eat, um, fruit.

Interestingly, our bodies, like those of other animals that have lost the ability to make vitamin C, have attempted to compensate by increasing dietary absorption of it. Animals that make their own vitamin C are typically very poor at absorbing it from their food because they just don't need it; humans, however, absorb dietary vitamin C at a much higher rate. But even though we have learned to eat food with ample vitamin C and even though our bodies are better at extracting these micronutrients from food, we have not managed to fully compensate for this malfunction. It's still a very poor design. In the days before fresh food from faraway places was readily available to people, scurvy was a common and often deadly disease.

Other essential vitamins can give us just as much trouble as vitamin C. Take vitamin D. The commonly ingested form of vitamin D is not fully active, which means that we can't use it until it's processed in the liver and kidney. The precursor of the vitamin is also generated in the skin, providing the individual gets enough sunlight, but it still needs to be processed into the active form. Without enough dietary vitamin D or enough sunlight, young humans can develop a disease called rickets, and older humans can develop osteoporosis. Rickets is extremely painful and leads to weak bones that break easily and heal slowly and, in severe cases, to stunted growth and skeletal deformities.

Both of these conditions involve brittle and deformed bones, which

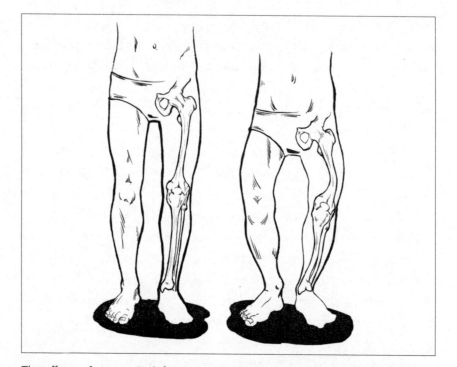

The effects of vitamin D deficiency on leg bones, a condition called rickets. Humans have trouble absorbing vitamin D from our diet; instead, our bodies require exposure to direct sunlight in order to synthesize it. If we fail to get enough vitamin D as children, the resulting skeletal deformities can last a lifetime.

can be extremely painful. Humans need calcium to keep bones strong, and we need vitamin D to help absorb calcium from food. We could eat all the calcium in the world, and none of it would be absorbed without sufficient vitamin D. (This is why vitamin D is commonly added to milk: it helps our bodies absorb the calcium that the milk contains.)

Rickets is a uniquely human disease for a variety of reasons. First of all, we're the only species that wears clothing and usually lives indoors. Both of these factors reduce the amount of sunlight exposure of the skin, thus crippling the ability to make the precursor of vitamin D. It could be argued that this is not a problem of poor design, per se, but it's certainly not *good* design. The complex multistep activation path for vi-

# B VITAMINS

| VITAMIN | ALIAS | FOOD SOURCE | EFFECTS OF DEFICIENCY |
|---|---|---|---|
| B$_1$ | Thiamine | Yeast, meat, cereals | Beriberi |
| B$_2$ | Riboflavin | Dairy, eggs, liver, legumes, leafy greens, mushrooms | Ariboflavinosis |
| B$_3$ | Niacin | Meat, fish, legumes, all grains except corn | Pellagra |
| B$_4$ | Choline* | | |
| B$_5$ | Pantothenic acid | Meat, dairy, legumes, whole grain cereals | Acne, paresthesia |
| B$_6$ | Pyridoxine | Fish, organ meats, root vegetables, cereals | Dermal, neurological disturbances |
| B$_7$ | Biotin | Most foods | Impaired neurological development |
| B$_8$ | Inositol* | | |
| B$_9$ | Folate | Leafy greens, fruits, nuts, seeds, beans, dairy, meat, seafood | Macrocytic anemia, birth defects |
| B$_{10}$** | PABA | | |
| B$_{11}$** | PHGA | | |
| B$_{12}$ | Cobalamin | Most animal-derived foods | Macrocytic anemia |

*Incomplete consensus on naming/identity. No longer considered a vitamin.*
**No longer considered a vitamin.*

The B vitamins and their deficiencies. While few if any wild animals have to contend with these deficiencies, they have been a significant blight for humans, particularly since the advent of farming and food processing.

tamin D is obnoxious enough, but requiring sunlight exposure to produce the precursor molecule adds another wrinkle — no pun intended — and is another way in which we can develop vitamin insufficiencies.

Second, due to modern lifestyles and diets, we don't always consume enough vitamin D. While it is always tempting to blame modern eating habits for dietary insufficiencies, that's probably not true in this case.

The innovations brought by civilization have *reduced* the incidence of rickets. To understand why, consider that, in order to get sufficient vitamin D in our diets, we need to eat at least some fish, meat, or eggs. Precivilization humans consumed few, if any, eggs. While meat and fish were staples, they were almost certainly not available on a steady basis. Prehistoric life was marked by periods of feast and famine, and we know from studying the bones of early humans that rickets and brittle bones were a constant problem. Not so for us modern humans in the developed world with our abundant sources of animal protein.

The domestication of animals for meat and eggs (roughly five thousand years ago in the Middle East and at different points elsewhere) mostly solved the problem of rickets. This is just one example of human ingenuity overcoming the design limitations of the human body—a theme we will encounter again and again in this book.

What about the many other vitamins listed on that bottle of multivitamins? Many of them fall into the family of B vitamins. There are eight different B vitamins, which often go by other names, such as niacin, biotin, riboflavin, and folic acid (or folate). Each of these vitamins is required for various chemical reactions throughout the body, and each has its own syndrome associated with insufficiency.

Among the most well-known B vitamin–deficiency syndromes results from not having enough vitamin $B_{12}$, also called cobalamin. This vitamin is familiar to long-term vegans because $B_{12}$ deficiency is a problem they invariably have to face; it leads to anemia. Humans cannot make their own vitamin $B_{12}$, and, since plants have no need for this vitamin, they do not produce it, so the only dietary sources of it are in meat, dairy, seafood, arthropods, other animal-derived foods, and vitamin supplements. Vegans, take note: you need these pills.

But what about vegetarian animals? There are many animals that eat only plants, but if plants don't have any $B_{12}$ and all animals need $B_{12}$ to survive, how do cows, sheep, horses, and the thousands of other herbivorous animals avoid anemia? The answer is that they make it—or, rather, the bacteria in their large intestines make $B_{12}$ for them.

You probably already know that the large intestines of mammals are chock-full of bacteria. Because bacteria are so much smaller than animal cells, there are more individual bacterial cells in your colon than there are human cells in your entire body. That's right—the bacteria that live in your body outnumber your own cells! They also do some important things for you. Vitamin K, for example, is made by the bacteria in the gut, and we simply absorb it from there. You don't need supplements or foods that contain it as long as you have the bacteria that produce it in your gut.

Just like vitamin K, vitamin $B_{12}$ is made by our intestinal bacteria —yet we need to get *more* vitamin $B_{12}$ in our diets. Why is that?

Here is the design flaw: Bacteria make $B_{12}$ in the *large* intestine, the colon, but we can't absorb it from there. We absorb $B_{12}$ in the *small* intestine, which comes *before* the large intestine in the flow of traffic within the digestive system. So the wonderful bacteria of the human gut are nice enough to provide $B_{12}$ for us, but the gut is so poorly designed that we send all of that $B_{12}$ to the toilet. (And, yes, in case you are wondering, you could eat your feces to get the $B_{12}$ you need, but I hope you will never be that desperate.) The bad plumbing job of our intestines has rendered $B_{12}$ an essential dietary vitamin for humans while all the millions of herbivore animals are blissfully unbothered by any need to find and eat this molecule.

The next most famous B vitamin–deficiency syndrome is beriberi, caused by a lack of vitamin $B_1$, also known as thiamine. Thiamine is required for a variety of chemical reactions in the body, the most important of which is converting carbohydrates and fats into usable energy. As a result of thiamine insufficiency, people can suffer nerve damage, muscle weakness, and heart failure.

Incredibly, despite this vitamin's importance, we cannot make it ourselves. Like $B_{12}$, vitamin $B_1$ must come from our diets. Also like $B_{12}$, $B_1$ can't be made by any animal. Only bacteria, most plants, and some fungi can make it, so at least we share this flaw with all our fellow animals. Except animals never get beriberi and humans have suffered massively from it. In fact, during the sixteenth and seventeenth centuries, it is estimated that beriberi was second only to smallpox in causing human death. Why only us?

The reason that other animals don't suffer from beriberi is that $B_1$

is abundant in a wide variety of plant foods found at the base of most food chains. In the oceans, many of the photosynthetic bacteria and protists found in plankton make $B_1$, and it proceeds up the food chain from there. Filter-feeding plankton-eaters, like the massive blue whale, get it directly, but carnivorous fish and mammals often eat things that eat things that eat plankton. In any case, $B_1$ makes the rounds. Same thing on land; many land plants are rich in $B_1$, meeting the dietary needs of herbivores, which are then eaten by carnivores, which are then eaten by apex predators, among which are humans, although we also eat plants, of course.

So why do humans struggle with beriberi when no other animals do? The answer seems to lie in how we prepare our food.

As humans invented and refined agriculture, they began to process foods in various ways to make them taste better and last longer without spoiling or becoming unpalatable. Often, these methods stripped many nutrients out of food.

For reasons that are not always understood, nutrients are not spread evenly throughout a plant. For example, the skins of potatoes and apples are where most of their vitamins A and C are located, so peeling them can rob them of most of their nutrients.

This can be seen acutely in the removal of rice husks. Unrefined rice, or brown rice, is rich in $B_1$. Refining raw rice, also called polishing, allows the rice to be dried and stored safely for years, and this agricultural innovation made a huge difference in preventing famine, especially in Asian populations, where rice is a staple. However, rice polishing removes essentially all the vitamin $B_1$. This was not a problem for the wealthy elite in Asian cultures, since the $B_1$-rich meat and vegetables that they ate supplemented the $B_1$-poor rice. However, for the vast majority of Asian people, beriberi was an endemic condition for thousands of years. It is still a concern in poor remote villages.

The scourge of beriberi may not technically be an example of poor human design since it has plagued us only since the dawn of civilization and is due to our own innovations. However, it *is* an example of how our evolutionary limitations can be exacerbated—or ameliorated—as we continue to develop as a species. Were it not for human innovations in agriculture and horticulture, civilization would not be possible in the first place. The same technology that led to high rates of beriberi allowed our species to advance past the hunter-gatherer lifestyle. Civilization enabled humans to lead healthier lives in a variety of ways, as evidenced by the explosion of the human population. Beriberi was a tradeoff our ancestors made unknowingly, because they didn't realize that their bodies could not produce a simple molecule required for the most basic chemical function: converting dietary calories into usable energy. So you could say that one cost of technology and civilization is beriberi.

To be sure, making our own vitamins is complicated and labor-intensive. Vitamins are complex organic molecules, many of them bearing a striking and distinct structure not closely related to other molecules'. To produce them, the body must have an elaborate pathway of enzyme-catalyzed chemical reactions. Each of those enzymes must be encoded by a gene. Those genes must be maintained, copied faithfully each time a cell divides, translated into proteins, and then regulated to match supply with demand. In the grand scheme of metabolism, the number of calories that an organism spends on synthesizing necessary vitamins is small, but it is not zero.

Given all of that, it is somewhat understandable why some organisms have forgone the making of their own vitamins and opted to obtain them from their diets instead. There is a certain logic to that; after all, why go to all of the trouble of making vitamin C when you already have it in your diet? Yet just because we don't always *need* to make some essential vitamins doesn't mean it's a good idea to relinquish the *ability* to make them; doing so would be terribly shortsighted, since humans would be stuck with that dietary requirement forever. Once a gene is broken, it is hard to unbreak it.

That logic does not apply to the essential amino acids, whose simple structures are very easy for cells to fashion. And yet we *still* can't make many of them either.

## Acid Tests

Amino acids are about as different from vitamins as two types of organic molecules can be. All organisms use twenty different kinds of amino acids to build proteins. Humans have tens of thousands of different proteins in their bodies, all of which are made with the same twenty building blocks. All twenty amino acids are structurally similar, each being a slight variation of another. To build these twenty amino acids, therefore, we don't need twenty separate pathways. Sometimes only a single chemical reaction is needed to change one amino acid into another. This is a far cry from the contortions the human body must go through to create different types of vitamins, and the uses for amino acids are much more varied than those for vitamins.

Nevertheless, we cannot make some of the amino acids for ourselves and must get them from our diets. In fact, nine of the twenty amino acids are called essential because we have lost the ability to manufacture them. I say that we have *lost* that ability because, as we look back through evolutionary time, we find ancestors who could make some or all of them. A wide swath of unrelated microorganism species (bacteria, archaea, fungi, and protists) can synthesize all twenty amino acids as well as the components needed for DNA, lipids, and complex carbohydrates. These extremely self-sufficient organisms can get by on just a simple carbon-based energy source, such as glucose, and a little organic nitrogen in the form of ammonia.

It's not just microorganisms that can make all their own amino acids either. Most plant species are capable of synthesizing all twenty amino acids. As a matter of fact, plants are even more self-reliant than most microorganisms because they can synthesize the energy source themselves, too, using energy from the sun. Given a simple balanced soil that contains organic nitrogen, many plants can live without any form of supplementation whatsoever. Plants don't *eat* anything. They make all of their own food internally. This remarkable self-sufficiency means that plants don't really require any other organisms, at least not for short-term day-to-day needs. This helps to explain how plants thrived on the dry land for a hundred million years, growing into impenetrably thick forests, before animals emerged from the oceans and began eating them.

Animals are the exact opposite of self-sufficient. They must constantly eat other living things in order to survive. They can eat plants, algae, or plankton, or they can eat other animals that eat them. Either way, animals must get all of their energy from organic molecules made by other living things since they cannot harvest solar energy themselves.

Since humans have to eat other living things anyway, we've gotten a bit lazy. While we eat plants and other animals mainly for their energy, consuming them also brings us all the proteins, fats, sugars, and even vitamins and minerals that those living things have in their bodies. We're not getting just energy when we eat, in other words; we're also getting various organic building blocks. This frees us from having to constantly make those molecules ourselves. If you are provided with a nice serving of the amino acid lysine, for example, every time you eat, why should you bother to spend energy making it?

Of course, each plant and animal has different amounts and combinations of amino acids. If we stop making lysine ourselves, we might be okay on a diet of fish and crabs (which are high in lysine), but a diet of berries and insects (low-lysine foods) would hurt us. That's the problem with discarding the ability to make certain nutrients. In order to save a few calories of energy, we lock ourselves into certain diets or lifestyles that we cannot change on pain of death. This is a dangerous game because the world is in a continuous state of flux. Every single geographic location and microenvironment has seen its share of upheavals, fluctuations, and catastrophes. The only constant in life is change.

Yet evolution has made these shortsighted tradeoffs in humans again and again. Our species has lost the ability to make nine of the twenty amino acids. Each loss is the result of at least one mutational event, usually more. Mutations happen to individuals randomly, of course; they become fixed in the population either by pure chance or because they offer some sort of distinct advantage. In the case of the mutations that destroyed our ability to make amino acids, it was probably a matter of chance.

When humans lost the ability to synthesize several amino acids, they gained nothing but the risk of debilitating, even deadly, dietary deficiencies—so how were these mutations not quickly eliminated when they occurred? Because our species' diets compensated for this loss, just like we saw with vitamin C. A diet with at least occasional meat or dairy consumption usually provides enough of each essential amino acid. Plant-based diets, however, need to be planned more carefully because different kinds of plants provide different ratios of the twenty amino acids. As a result, variety is the easiest way for vegetarians and vegans to ensure that they get enough of all the amino acids they need.

In the developed world, it is not difficult for a vegan to acquire all the nine essential amino acids. A single serving of rice and beans can give a whole day's supply, provided that the rice is unrefined and the beans are of the black, red, or kidney variety. Further, chickpeas, also called garbanzo beans, contain large quantities of all nine essential amino acids all by themselves, as do quinoa and a few other so-called superfoods.

However, among the poor, especially in developing countries, a varied diet is not always an option. There are billions of humans who subsist on extremely simple diets consisting of just a few staples, and those staples often do not provide enough of some of the essential amino acids, especially lysine. In some remote Chinese villages, the poorest of the poor will live on nothing but rice and the occasional scrap of meat, egg, or bean curd. In the poorest parts of Africa, the most destitute subsist on diets composed almost entirely of wheat products, and even those become scarce during famines. Unsurprisingly, given examples like these, protein deficiency is the single most life-threatening dietary problem in the developing world. This problem stems directly from our species' inability to make certain amino acids.

The problem of amino acid deficiency is not unique to the modern world by any means. Preindustrial humanity probably dealt with protein and amino acid insufficiency on a regular basis. Sure, large hunted animals such as mammoths provided protein and amino acids aplenty. However, living off big game in the era before refrigeration meant humans had to endure alternating periods of feast and famine. Droughts, forest fires, superstorms, and ice ages led to long stretches of difficult conditions, and starvation was a constant threat. The human inability to synthesize such basic things as amino acids certainly exacerbated those crises and made surviving on whatever was available that much harder. During a famine, it's not the lack of calories that is the ultimate cause of death; it's the lack of proteins and the essential amino acids they provide.

Amino acids are not the only basic biomolecules that humans and other animals have lost the ability to synthesize. Two other examples come from a group of molecules called fatty acids. These long hydrocarbons are the building blocks for fats and other lipids that the body needs, such as phospholipids, which help form the membranes that surround every single cell. It is hard to think of a more essential structure than the cell membrane. Yet one of the two fatty acids that we cannot produce (both of which have tongue-twister names) is linoleic acid, which forms part of the cell membrane. The other one, alpha-linolenic acid, is used to help regulate inflammation, another hugely important internal process.

Luckily for us, modern human diets have provided these two essential fatty acids in sufficient quantities in the form of seeds, fish, and various vegetable oils. Fortunately, too, several studies have shown that frequent consumption of these fatty acids leads to improved cardiovascular health. But we weren't always so lucky. In prehistory, especially in the era before agriculture, human diets tended to be much simpler. Roving bands ate what they could find, doing their best to follow the food. Most of the time, these fatty acids were probably available, but there can be little doubt that periods of deficiencies existed as well. Sometimes grass, bugs, leaves, and the occasional berry were all that could be found. Just as with essential amino acids that we cannot synthesize ourselves, losing the supply of two important fatty acids would make any food crisis that much worse.

What is most maddening about these two fatty acids is that they can be made pretty easily. Our cells can synthesize a whole host of lipid molecules, many of them quite a bit more complex than linoleic acid and alpha-linolenic acid. In fact, we make many very elaborate lipids *from* these simple ones—yet we cannot make these two themselves. The enzymes necessary to produce these particular fatty acids exist in many organisms on earth, but humans aren't one of them.

The human body, like the bodies of all animals, takes in plant or animal tissue, mashes it up, absorbs the small constituents, and uses the little bits to build its own molecules, cells, and tissues. However, there are gaps in this scheme. There are several molecules that are crucial for human health that we are incapable of making, so we have no choice but to seek them out in our food. The fact that we need to find these essential nutrients places restrictions on where and how humans can live. And that's just organic nutrients. The human body is also terrible at obtaining the inorganic kind—known as minerals—even when they are right there in what we eat.

## Heavy-Metal Machines

For squishy, water-based creatures, we humans sure do need a lot of metal in our diets. There are all kinds of metals—known as essential minerals—that we have to eat. Metal ions are single atoms, not complex molecules, and they cannot be synthesized by any living thing. They must be ingested in food or water, and the list of ions that are essential for us includes cobalt, copper, iron, chromium, nickel, zinc, and molybdenum. Even magnesium, potassium, calcium, and sodium are technically metals, and we need substantial amounts of these minerals daily too.

We don't think of these minerals as metallic because we don't consume or use them in their elemental forms. Instead, cells use metals in their water-soluble, ionized forms. To appreciate the stark difference, consider sodium.

Sodium, as it appears on the periodic table in its elemental form, is a metal so reactive that it catches fire if it comes in contact with water. It's highly toxic; a tiny amount could kill a large animal. However, when we remove a single electron from a sodium atom, turning it into an ion, it has completely different properties. Ionized sodium is more than simply harmless; it is essential for all living cells. It combines with chloride ions to form table salt. For all intents and purposes, elemental sodium (Na) is a completely different substance than ionized sodium ($Na^+$).

While sodium and potassium are inarguably among the most important of the metal ions (in the sense that no cells can function without them), humans almost never have a chronic lack of these minerals in their diets. All living things have these two ions in relative abundance, so whether you're paleo, vegan, or something in between, you will get the sodium and potassium that you need. *Acute* deficiencies of sodium or potassium can be an urgent problem, but they are usually the result of physiological dysfunction, fasting, excessive dehydration, or some other short-term insult.

For other essential ions, the story is different. If you aren't eating 'em, you aren't getting enough of them, and you'll suffer chronic illness as a result. Inadequate calcium intake, for example, is a problem throughout the world, affecting both rich and poor. Calcium insufficiency is one of the most frustrating dietary problems from a design standpoint because lack of calcium stems from our species' poor ability to absorb it rather than from not having enough of it in our food. We all *eat* plenty of calcium; we just aren't very good at *extracting* it from food. As already mentioned, vitamin D is required for calcium absorption, so if you're deficient in vitamin D, all the dietary calcium in the world cannot help you because it will proceed right through your gut undisturbed.

Even if we have plenty of vitamin D, we still aren't very good at absorbing calcium, and we get worse and worse at it as we age. While infants can absorb a respectable 60 percent of the calcium they consume, adults can hope to absorb only around 20 percent, and by retirement age, that drops to 10 percent or even lower. Our intestines are so bad at extracting calcium from food that our bodies are forced to extract it from our bones instead—a strategy with devastating consequences. Without constant calcium and vitamin D supplementation, most people would develop the brittle bones of osteoporosis in their golden years.

In prehistoric times, few humans lived beyond thirty or forty, so you might think that calcium deficiency was not such a problem for our ancestors. Yet even so, the majority of ancient skeletal remains show the telltale signs of calcium and vitamin D deficiency, and they appear more drastically—and in younger people—than we typically see today.

So osteoporosis and the calcium shortages that can create it are definitely not new problems. Neither is the difficulty humans often encounter with getting enough of another vital mineral: iron.

Iron is the most abundant transition metal (the class of metals occupying the huge center section of the periodic table and known for conducting electricity well) in our bodies and in the earth. As with the other metals, we make use of ionized iron atoms, not the elemental metal form. Most of that elemental stuff sank to the core of the earth shortly after it formed; what we have here on the surface is mostly the ions lacking one, two, or three electrons. In fact, the ease with which iron can switch among these different ionized states is the secret behind its special utility in our cells.

The most commonly known role of iron is in the functioning of hemoglobin, the protein that transports oxygen throughout our bodies. Red blood cells are absolutely packed with this protein, each molecule of which needs four iron atoms. In fact, the iron atoms in hemoglobin are what give it its characteristic red color (which means your blood and the surface of Mars have more in common than you might think). Iron is also vital for other crucial functions, including the harvesting of energy from food.

Despite the fact that there is plenty of iron in our bodies, our environment, our earth, and our solar system, deficiencies in iron are among the most common diet-related ailments in humans. In fact, according to the Centers for Disease Control and Prevention (known as the CDC) and the World Health Organization (WHO), iron deficiency is the single most common nutritional deficiency in the United States and worldwide. That iron deficiency is pandemic in a world filled with iron is paradoxical, to say the least.

The most acute problem caused by iron insufficiency is anemia, a word that loosely translates to "not enough blood." Because iron is central to the hemoglobin molecule and hemoglobin is central to red blood cell structure and function, low iron levels impair the body's ability to make blood cells. The WHO estimates that 50 percent of pregnant women and 40 percent of preschool children are anemic due to iron deficiency. Current estimates are that two billion of the world's seven billion people are at least mildly anemic. Millions die from the deficiency each year.

Once again, poor design is mostly to blame for the body's problems. To start with, the human gastrointestinal tract is terrible at extracting iron from plant sources.

Plant- and animal-derived iron are structurally different things. In animals, iron is generally found in blood and muscle tissue, and it's easy enough to process; humans usually have little trouble extracting iron from a nice hunk of steak. The iron in plants, however, is embedded in protein complexes that are much harder for the human gut to rip apart, and so they remain in the gastrointestinal tract and end up as waste, making iron consumption another concern for vegetarians. In this respect, humans are worse off than most animals. The majority of the creatures on earth are mostly or completely vegetarian, yet their intestines do just fine in processing iron.

Additionally, there are many quirks about iron consumption that can further reduce absorption of it. For instance, we absorb iron best when it comes together with something else we readily absorb—for example, vitamin C. Vegetarians use this trick to boost their iron absorption. By combining sources of iron with sources of vitamin C, they can ensure that their bodies are better able to absorb both. A large dose of vitamin C can increase iron absorption sixfold. Unfortunately, the opposite is also true; a diet poor in vitamin C makes iron absorption more difficult, often leading to the double whammy of scurvy and anemia. Just imagine that combination. It's bad enough that you are pale and lethargic, but you could also lose muscle tone and begin bleeding internally. Vegetarians in developed countries avoid this lethal trap because they have access to many foods that are high in both iron and vitamin C, such as broccoli, spinach, and bok choy. Poor people in the developing world are usually less fortunate, however, as those key foods are often precious and strictly seasonal.

As if getting enough iron weren't hard enough already, there are several other food molecules that actually interfere with iron absorption, particularly the iron in plants. Foods such as legumes, nuts, and berries

—which we're told to eat plenty of—contain polyphenols, which can reduce our ability to extract and absorb iron. Similarly, whole grains, nuts, and seeds are high in phytic acid, which tends to prevent iron from being absorbed by the small intestine. These complications are especially problematic for the two billion people on the planet who are at risk of anemia due to poverty, those for whom meat, and the iron in it, is a rare treat. Their diets tend to be high in the very foods that make iron extraction from plant sources even more difficult. While eating a varied diet is a good strategy to acquire all the elements we need, including iron, it must be *carefully* varied, in such a way that iron-rich foods are not paired with those that prevent iron extraction.

Another dietary component that interferes with iron absorption is calcium, which can reduce iron absorption by up to 60 percent. Thus, foods rich in calcium, such as dairy, leafy greens, and beans, should be consumed separately from foods rich in iron in order to maximize absorption, especially if the source of the precious iron in question is plant-based. If you go to the trouble of eating iron-rich foods but pair them with calcium-rich foods, you've negated your efforts. It's not enough to eat the right foods to meet our exacting dietary needs; we must eat those foods in the correct combinations. It's no wonder so many of us opt for a multivitamin instead.

Iron deficiency is yet another example of how our species' prehistoric diet was even more insufficient than our modern diet. Although meat and fish were likely staples of the early human diet, their availability waxed and waned both seasonally and through long periods of feast or famine, and proteins were particularly hard to come by in landlocked communities that relied on meat alone. Before agriculture, the available food plants were nothing like the food people are accustomed to eating now. Fruits were tiny and bland, vegetables were bitter and mealy, nuts were hard and tasteless, and grains were tough and fibrous. Worse, plants that reduced iron absorption were more common than plants that provided iron.

While it's not all that hard to get enough iron from a vegetarian diet *nowadays*, it would have been nearly impossible during the Stone Age. Most prehistoric humans would have suffered from severe anemia whenever meat was scarce. This is at least part of the reason why migrations of preagricultural human communities largely followed coastlines or other bodies of water: fish were a more reliable source of iron than meat.

You may be wondering how, if anemia is such a lethal and constant danger, humans survived at all. We almost didn't. Our species teetered on the verge of extinction throughout much of prehistory. Over the past two million years, several species of hominids have come and gone, with all but one lineage ending in extinction. At certain points in our species' long journey, our ancestors were so few in number that they surely would have been classified as endangered by today's standards. What's more, none of these branches of hominids was more cognitively advanced than any other until very recently, so modern humans can't credit their big brains with surviving each and every brush with extinction; it was probably blind luck that saved our forebears in more than a few instances. These near demises had a variety of causes, but iron-deficiency anemia was almost certainly among them.

To add insult to injury, humans appear to be unique in the struggle to maintain healthy levels of iron. Neither rampant anemia nor iron deficiency has been documented in any successful species besides ours.

How, then, do other animals deal with the challenge of acquiring sufficient iron? After all, the need for this essential mineral is not uniquely human, and no other animals can produce it either. Surely evolution has devised solutions to this challenge somewhere along the line, even if not in humans?

The answer is complicated. First of all, aquatic animals, whether fish, amphibians, birds, mammals, or invertebrates, do not have the same challenge in getting iron because iron ions are plentiful in both seawater and fresh water. These animals still need to extract iron ions from the water, of course, but finding it is a nonissue. Similarly, iron is abundant in rocks and soil, so plants get it easily.

It appears that herbivorous or mostly vegetarian animals are either better at incorporating abundant iron sources into their diets or better at extracting it from their food than we are. When these species experience famine, displacement, or other stressors, sure, iron deficiency is common, but that is an *effect,* not a cause. Humans are the only animals that seem to suffer from iron deficiency even when they're doing fine otherwise.

The frustrating part is that we don't completely understand why it's so hard for us to get enough iron. Why are humans so bad at extracting iron from plant sources? Why are we so sensitive to the inadvertent pairing of iron-rich foods with those that inhibit the ability to extract it? These do seem to be uniquely human problems. It is possible that humans suffered one or more mutations in the genes that are responsible for iron absorption, and it just didn't matter at the time because our ancestors had rich sources of animal-derived iron in their diets, probably fish or big game. That's a plausible hypothesis, although it has not yet been proven.

Deficiencies of other heavy metals are much more rare than iron deficiency, mostly because we need so little of these other minerals. We need only the tiniest amounts of copper, zinc, cobalt, nickel, manganese, molybdenum, and a few others. In some cases, we can go for months or years at a time without ingesting those metals, relying solely on our internal reserves.

Nevertheless, trace amounts of these heavy metals are crucial, and a diet completely devoid of them would eventually be lethal. Was it an evolutionary error that made it so hard for humans to absorb them, or was it just a failure to adapt to this challenge? Is there a difference? There are plenty of microorganisms that simply have no need for many of these elements. In fact, there is no single trace metal that is required by all organisms. Put another way, for every one of these elements, there are organisms that have engineered their own molecules to perform those elements' jobs. Humans haven't done much of that and so we require a broad variety of trace metal ions.

## Coda: Gorged to Death

For decades, the United States and other developed countries have been flooded with diet books. This reflects an ominous trend. Starvation used to be a serious threat to all humans; now, obesity is replacing it as a scourge in many parts of the world.

This flows directly from the shortsighted way that evolution programmed our bodies. As many of these books note, we are hardwired for obesity. Yet most popular explanations of how and why things have gone wrong miss the evolutionary lesson at the heart of this growing problem.

Virtually every human being loves to eat. Most people are constantly craving food, whether they are really hungry or not, and cravings are usually for high-fat, high-sugar foods. But most of the foods that supply essential vitamins and minerals — from fruits to fish to leafy greens — are not high in sugar or fat. (When was the last time you had an intense craving for broccoli?) So why do our instincts drive us to high-calorie foods no matter how well fed we are?

Because obesity has been on a steep and steady rise and was not a major health concern until the past century or two, it is tempting to consider it a problem of modernity, not biology. But while it's true that modern lifestyles and eating habits are to blame for the current rate of obesity in the developed world, that is putting the cart before the horse. People don't eat too much just because they *can*. They eat too much because they were *designed* to. The question is, Why?

Humans are not unique in being gluttonous. If you have dogs or cats, you have undoubtedly noticed that their appetites seem to be insatiable. They always want more treats, more scraps, more food, and they will beg more persistently for rich and savory foods than for, say, salad. In fact, our companion animals are just as prone to obesity as we are. If we are not careful about how much food we give them, they will become overweight quickly.

Scientists know that this is true for laboratory animals as well. Whether they are fish, frogs, mice, rats, monkeys, or rabbits, their food has to be limited or they will become overweight. The same is true at zoos. Animal handlers and veterinarians constantly monitor the weight and food rations of the captive animals so their health won't suffer due to overconsumption.

The take-home point here is that all animals, including and especially humans, will become morbidly obese if left purely to their own devices. This is in direct contrast to what we see with animals in the wild, where obesity is rare, to say the least. Animals living in their natural habitats are almost always tight and trim — skinny, even.

It was once thought that the artificial environments of zoos, laboratories, and human homes were to blame for animals' obesity. After all, animals have spent millions of years adapting to their wild natural habitats, and the artificial ones are just no substitute. Perhaps the stress of captivity causes nervous overeating. Or maybe the comparatively sedentary lifestyle throws the metabolic dynamics out of balance.

While these are reasonable hypotheses, they've been tested over the years and don't seem to be the main explanations for captive animals' obesity. Captive animals that exercise still need their food rationed. They will still become obese if they're provided with too much.

So why don't we find obese animals in the wild? The answer — a rather disturbing one — is that most wild animals are teetering on the edge of starvation pretty much all the time. They live their lives in a state of constant hunger. Even animals that hibernate and spend half the year bingeing are still relentlessly hungry. Surviving in the wild is a brutal business and a perpetual struggle. Different species of animals are in constant competition with one another for scarce resources, and there is simply never enough food. This scarcity of food is a biological constant for all animals—except modern humans.

For much of the twentieth century, it was thought that modern lifestyles and conveniences were to blame for the emergence of obesity. Desk jobs had begun to replace manual labor, and radio and television were replacing sports and other forms of physical recreation. The thinking was that previous generations had been much more physically active in both their livelihoods and their entertainment. The increasingly sedentary lifestyle and the transition away from physical toil was to blame for bulging waistlines. By this reasoning, obesity is not the result of a design flaw; it's the result of poor lifestyle.

Although this may seem to make sense, it is not the whole story. For one thing, people who do make their living through physical toil are not in any way immune from obesity. The opposite is true, in fact, as both obesity and physical labor correlate with lower income. Second, children who spend more time engaging in physical play than indoor play are no less likely to develop obesity as adults. Again, the opposite tends to be true: people who are active athletes throughout their childhood, adolescence, and even into adulthood are *more* likely to be obese in their thirties, forties, and fifties, particularly when their physical activity wanes. It isn't lifestyle, it's the overconsumption of calorie-rich foods that seems to be the main cause of obesity.

This explains why, unfortunately, exercise alone rarely leads to long-term weight loss. In fact, exercise may do more harm than good. Intense exercise leads to intense hunger, which in turn leads to poor diet choices and chips away at the mental resolve to lose weight. Each time someone slips on a diet, he gets closer to just giving up altogether.

The hard truth is that humans in the developed world are surrounded with high-calorie foods that they are ill equipped to resist. For most of our species' history, this just hasn't been something anyone needed to worry about. Until the past couple of hundred years, most people didn't have access to diets rich in meat and sweets. It was the industrial revolution that began to bring rich diets to the masses. Before that, stoutness in a man and plumpness in a woman were signs of wealth, power, and privilege, and the commoners were, like animals in the wild, prone to constant hunger.

Overeating was a fine strategy when it wasn't possible to do very often. But when people can pack it in three or four times a day, day after day, their feeble willpower doesn't stand a chance of moderating intake to prevent unhealthy weight gain. Human psychology is no match for human physiology. This is why people too often treat every meal as if it were the last one before a long winter, as if they're gorging themselves in anticipation of barely being able to find any food at all.

It gets worse. As recent studies have shown, our bodies adjust our metabolic rates so we gain weight easily and lose weight only with difficulty. Those who have struggled with their weight will tell you that weeks of dieting and exercise often result in negligible weight loss, while a weekend calorie binge can pile on a few pounds almost instantly. Thus, obesity and type 2 diabetes are the quintessential evolutionary mismatch diseases, conditions that directly result from humans living in a very different environment than the one they evolved in.*

Thanks to modern food supplies, people in the developed world will probably never need to worry about scurvy, beriberi, rickets, or pellagra. Obesity, however, will be a constant challenge to their willpower and habits. There is no quick fix. This fatalistic truth is reminiscent of the next categories of flaws we will explore—the flaws in our genomes.

---

* I highly recommend Dan Lieberman's book *The Story of the Human Body*, which explains at great length the many ways in which the mismatch between our current environment and our former one leads to illness and disease.

# 3

# Junk in the Genome

*Why humans have almost as many broken, nonfunctional genes as functioning ones; why our DNA contains millions of virus carcasses from past infections; why a bizarre self-copying piece of DNA makes up over 10 percent of the genome; and more*

You may have heard that humans use only 10 percent of their brains. This is a total myth; humans use every lobe, fold, and nook of their neural tissues. While some regions specialize in certain functions—speech, for instance, or movement—and rev up their activity when performing them, the whole brain is active pretty much all of the time. There is no part of the brain, no matter how tiny, that can be deactivated or removed without serious consequences.

Human DNA, however, is a different case altogether. There are vast expanses of our genomes—the entirety of the DNA we carry in each of our cells—that do not have any detectable function. This unused genetic material was once referred to as *junk DNA* to reflect its presumed uselessness, although this term has fallen out of favor with some scientists, as they have discovered functions for some parts of this "junk." Indeed, it may very well turn out that a large portion of so-called junk DNA actually serves some purpose.

Regardless of how much junk our genomes contain, however, it is undeniable that we all carry around a whole lot of nonfunctional DNA. This chapter tells the story of that *true* genetic junk: the broken genes, viral byproducts, pointless copies, and worthless code that clutter our cells.

Before we proceed, it is worth pausing for a quick refresher about the basics of human genetics. Almost every one of your cells, whether it's a skin cell, a muscle cell, a neuron, or any other type of cell, has within it a core structure called a nucleus that contains a copy of your entire genetic blueprint. This blueprint—much of which is illegible, as we'll see—is your genome, and it is composed of a type of molecule called deoxyribonucleic acid, better known as DNA.

DNA is a linear double molecule that looks like a very long twisted ladder, and the genetic information it contains is written in pairings of other, smaller molecules called nucleotides. Think of these nucleotide pairs as the rungs of this metaphorical ladder. Every rung has two halves, each of which is a nucleotide molecule that's attached to one of the two sides of the ladder. These nucleotides come in four flavors, abbreviated A, C, G, and T; A can pair only with T, and C can pair only with G. These are known as base pairs, and they are what make DNA such an incredibly effective carrier of genetic information.

If you look along one side of the ladder that is your DNA, you can see any combination of the four nucleotide letters. Let's say you are looking at five rungs, and on one side you see the letters A, C, G, A, and T. Because the rung pairs can only put A with T and C with G, you can be confident that if you moved around to the opposite side of the ladder and looked at the other half of these same rungs, you would see a mirror image of the code on the other side (with the sequence reversed): A, T, C, G, and T.

This is a simple but ingenious form of information coding, especially because it makes it very easy for genetic material to be copied again and again. After all, you could rip your entire, massively long ladder down the middle, splitting every rung in half, and each half would essentially contain the same information. This is precisely what the cell does in order to copy the DNA molecule prior to cell division, the basic process by which the body replaces old cells with new ones. So DNA's ability to copy itself is not only a miraculous feat of evolutionary engineering but also the basis of our very existence.

So far, so good; DNA is a wonder of nature. But here's where it starts to look less wondrous. The ladder of DNA that makes up your genome has billions of rungs, 2.3 billion in total, composed of 4.6 billion letters. And a lot of those rungs are, for lack of a better word, unusable. Some of them are pure repetitive nonsense, like someone was pounding on a computer keyboard for hours, while other bits were formerly useful but became damaged and were never repaired.

If you read along the entirety of either side of the ladder of your DNA, you'd notice something strange. Your genes, those sections of the code that can actually accomplish something (cause the irises of your eyes to take on a certain color, say, or direct you to develop a nervous system), are only about 9,000 letters long on average, and you have only around 23,000 of these genes in total. That might seem like a lot, but in truth it's only a few hundred million letters of DNA—a couple hundred million rungs out of the 2.3 billion that make up your body's genome.

What are all those other rungs doing if they're not part of your genes? The short answer is: nothing.

To understand how this could be, let's adopt a new analogy. Let's call genes *words,* a string of letters of DNA that add up to something meaningful. In the "book" that is your genome, the spaces between these words are filled with incredibly long stretches of gibberish. All told, only 3 percent of the letters in your DNA are part of words; most of the remaining 97 percent is gobbledygook.

You don't have just *one* long ladder of DNA. Each cell has forty-six of them, called chromosomes, and they can actually be seen under a regular microscope in the moments that a cell is dividing. (The exceptions are sperm and egg cells, which have only twenty-three chromosomes each.) When cells aren't dividing, however, all of the chromosomes are relaxed and mushed in together like a big bowl of forty-six tangled spaghetti noodles. The chromosomes vary in length, from chromosome 1, with two hundred and fifty million of these rungs, to chromosome 21, with just forty-eight million rungs.

While some chromosomes exhibit a fairly high ratio of useful DNA to junk, others are littered with repetitive, unused DNA. For example, chromosome 19 is fairly compact, with over fourteen hundred genes spread over fifty-nine million letters. On the other extreme is chromosome 4, which is three times larger than 19 but has around half the number of genes. These functional genes are few and far between, like small islands surrounded by vast, empty oceans.

In this regard, the human genome mirrors that of other mammals, and all mammals have around the same number of genes, roughly twenty-three thousand. Although some mammals have as few as twenty thousand and others have as many as twenty-five thousand, this is still a relatively tight range — and one that is particularly surprising, given that the mammalian lineage is more than two hundred and fifty million years old. It is quite remarkable that, even though humans have been evolving separately from some mammals for over a quarter of a billion years, we all have a similar number of functional genes. In fact, humans have roughly the same number of genes as microscopic roundworms, which have no real tissues or organs. Just saying.

While relatively sparse, operational genes do a lot of work. They each make proteins by ripping the DNA molecule's ladder down the middle and exposing all the letter nucleotides of either side. The stretch of letters that make a gene can be copied into something called mRNA (*m* for "messenger," and *RNA* for "ribonucleic acid"), which in turn makes a protein that can travel around cells and help everything happen — things like growing and staying alive.

These twenty-three thousand genes that collectively make up 3 percent of the genome are a wonder of nature. Most of the other 97 percent of human DNA is more of a blunder — it does not seem to do very much. Some of it, indeed, is actually harmful.

The entire genome — functional or not — gets copied every time a cell divides. This consumes cellular energy and requires time, energy, and chemical resources. The best estimates are that a human body experiences at least $1 \times 10^{11}$ cell divisions each day. That's over a million cell divisions per *second.* Each one of those divisions involves copying the entire genome, junk and all. You expend a few dietary calories every single day just to copy your largely useless DNA.

Oddly, cells meticulously spot-check this junk DNA for errors. Each time a cell copies this irrelevant DNA, it engages the same proofreading and repair mechanisms that it does when it copies the most important genes in the genome. No region is ignored; none gets special attention. This is perplexing because a replication error in a stretch of gibberish DNA is inconsequential, while the same mutation in a gene can be lethal, as we'll see shortly. The machinery for copying and editing DNA can't seem to differentiate between the two types — genes and gibberish — any more than a chimpanzee can differentiate between a poem by a preschooler and one by Maya Angelou.

We are in an exciting new era of biomedical research. Scientists can now read the entire sequence of someone's genome, all *4.6 billion* letters spread across the forty-six chromosomes, in a process that takes just a couple of weeks and costs about a thousand dollars. (The first complete sequencing of a human genome took over a decade to finish and cost nearly three hundred *million* dollars.) But although we are finding many surprising new functions for *some* regions of DNA formerly referred to as junk, these are still overshadowed by the nonfunctional junk. What's more, even the apparently functional junk probably *began* as pure junk.* Given all of the nonsense encoded in the human genome, it's amazing we've turned out as well as we have.

Although the enormous tangled masses of useless DNA in the genome may be the biggest flaw of all, even the functional parts of it — the genes — are rife with errors. These errors generally come from mutations, which are what we call the changes that get made to a DNA sequence. There are two common ways that the genome can experience sudden changes. (Three if you count retroviral infections, but we'll talk about those soon enough.) One is through damage to the DNA molecule itself. This can occur due to radiation, UV light, or harmful chemicals called mutagens, such as those abundant in cigarette smoke. (Mutagens are often called carcinogens due to the mutations' tendency to cause cancer.)

A second way that the genome can experience changes is through copying errors made when DNA is duplicated in preparation for cell division. Each cell has 4.6 billion letters of DNA code, and each day the average person experiences somewhere around $1\times10^{11}$ cell divisions, so that is literally $10^{20}$ (100,000,000,000,000,000,000 or 100 quintillion) chances a day for a cell to make a mistake while copying DNA. Cells are terrific copyeditors, making fewer than one mistake in a million letters and immediately catching and correcting

---

* In 2012, the massive genome-exploring project called ENCODE made a big splash by claiming that up to 80 percent of the human genome was functional. This claim has been soundly refuted, owing partially to methodological concerns but mostly to researchers' unscientific criteria for declaring a part of the genome functional. This has led many scientists to revisit and defend the use of the term *junk* to describe nonfunctional DNA. See Graur et al., "On the Immortality of Television Sets," *Genome Biology and Evolution* 5 (2013): 578–90, for an excellent explanation of the flawed reasoning of the ENCODE claims.

99.9 percent of those rare errors. But even with that incredibly low error rate, with so many chances to make a mistake, sometimes mistakes do happen and don't get corrected. Those become mutations. In fact, every day, you experience millions of mutations throughout your body.

Fortunately, most of those mutations occur in unimportant regions of the DNA so they don't really matter. Further, mutations in cells that are not sperm or egg cells have no real consequences for evolution because they don't get passed on. Only the DNA in the so-called germ cells contributes to the next generation.

However, both copying errors and DNA damage can and do occur in important regions of the genome of sperm or egg cells. When this happens, these mutations will probably not affect that individual so much as his or her children. These are thus called heritable mutations and are the basis of all the evolutionary changes and adaptations in living things. But it's not all happy accidents when it comes to heritable mutations. While most are inconsequential (considering how much of the genome doesn't do anything anyway), many mutations are harmful because they disrupt the function of a gene.

The poor offspring that inherit a gene mutation from their father or mother are almost always worse off for it. That's what natural selection is all about—keeping the gene pool clean—but sometimes the harm that a mutation brings is not immediate. If a mutation causes a human or animal no short-term loss in health or fertility, it won't necessarily be eliminated. It could even spread throughout a population. If this mutation causes harm only far down the road, natural selection is powerless to immediately stop it.

This is evolution's blind spot, and its ramifications can be seen throughout our species—and deep inside each of us. For the human genome contains thousands of scars from harmful mutations that natural selection failed to notice until it was too late.

## Broken Genes

Of the useless DNA in the human genome, one particular kind stands out: pseudogenes. These stretches of genetic code look like genes but do not function as such. They are the evolutionary remnants of once-functional genes that became mutated beyond repair at some point in our species' deep past.

We saw one such pseudogene in the previous chapter, the *GULO* pseudogene, which in its functional form allows nearly all nonprimate animals to synthesize their own vitamin C. In some common ancestor of all living primates, the *GULO* gene was damaged by random mutation. Because this ancestor happened to have a diet rich in vitamin C, the mutation didn't cause any harm to that individual. However, since it was passed on to all primates, they — we — are subject to the horrors of scurvy.

You might wonder why nature wasn't able to fix this problem the same way it created it: through mutation. That'd be nice, but it's nearly impossible. A mutation is like a lightning strike, a random error in the process of copying 4.6 billion letters of DNA. The odds of lightning striking the same place twice are so infinitesimally tiny as to be nonexistent. What's more, it's exceedingly unlikely that a mutation will *fix* a broken gene because, following the initial damaging mutation, the gene will soon rack up additional mutations. If the first mutation doesn't kill or harm the bearer, future mutations won't either. These mutations will not be eliminated by natural selection.

This is why, over the scale of evolutionary time, the mutation rate of pseudogenes is dramatically higher than that of functional genes. Mutations in functional genes don't usually persist across generations; typically, the lightning strike will cause such harm to the cell or organism in which it occurs that the individual will be less likely to reproduce successfully, thereby limiting the spread of the faulty genetic material. Pseudogenes, however, are free to accumulate mutations without harming the entity that carries them—and that's exactly what happens. The pseudogenes get passed on and on, continuing their deterioration down through the generations, and it doesn't take long before the gene is mutated beyond all hope of repair.

That's what happened to the human *GULO* gene. Compared to the functional version that most other animals have, our *GULO* gene is littered with hundreds of mutations. Yet it is still easily recognizable. Our *GULO* gene is more than 85 percent identical to the DNA sequence of the functional *GULO* gene found in carnivores such as dogs and cats. It's mostly all there, just sitting useless, like a car rusting in a junkyard —except that humans have continually refashioned this rusty old gene, billions of times every day, ever since the gene was initially broken tens of millions of years ago.

Thanks to scurvy, *GULO* may be humans' most famous pseudogene —but it is hardly the only one. We humans have quite a few broken genes lurking in our genomes—actually, well over a few; more than a hundred or even a thousand. Scientists estimate that the human genome contains the intact remnants of nearly *twenty thousand* pseudogenes. That's almost as many broken genes as functional ones.

To be fair, the majority of these pseudogenes are the result of accidental gene duplications. This explains why the disrupting mutations and subsequent death of the gene didn't have any deleterious effects on the individual—these were extra copies of genes anyway. Their function was redundant with that of other genes, so losing these genes didn't put anyone at a disadvantage. Of course, it's still pointless to keep them around and duplicate them constantly. Pointless, and a waste of energy —but not directly harmful.

But when the only copy of this or that functional gene gets broken by a mutation and becomes a pseudogene, it can really hurt. Besides *GULO* (and its gift of scurvy), another pseudogene whose breakage had a deleterious impact on our species' health is a gene that once helped our ancestors fight infections. This gene created theta defensin, a protein still found in most Old World monkeys, New World monkeys, and even one of our fellow apes, the orangutan. However, in an ancestor common to humans and our African ape relatives, gorillas and chimpanzees, it was deactivated, then mutated beyond repair. Without a working version of it, humans are more susceptible to infections than our more distant primate cousins.

Admittedly, we have probably evolved other defenses to take the place of these theta defensin proteins—but not enough, it seems. For instance, cells that lack theta proteins appear more susceptible to HIV infection. We really could have used these proteins in the late 1970s and 1980s as HIV ravaged human populations around the globe. Were it not for this broken gene, the AIDS crisis might never have happened, or at least it might not have been so widespread and so deadly.

Pseudogenes are a lesson in the cruel habits of nature, which takes no thought for the morrow. Mutations are random, and natural selection operates merely from one generation to the next. Evolution, however, operates on very long timescales. We are the long-term products of short-term actions. Evolution is not—indeed, cannot be—goal-oriented. Natural selection is affected only by immediate or very short-term consequences. It is blind to long-term ones. When *GULO* or the gene that produces theta defensin die by mutation, natural selection can protect the species only if the lethal effects are felt immediately. If the bearers of the mutation continue to flourish and pass it down to their progeny, evolution is powerless to stop it. The death of the *GULO* gene probably had no consequence at all for the first primate that endured it, yet his or her distant progeny suffer still, several tens of millions of years later.

The *GULO* gene and the theta defensin gene are not unique in having suffered such debilitating mutations. Every single one of our other twenty-three thousand genes has been, and still is being, struck and killed by the lightning bolts of mutation. The only reason humans haven't lost more genes to mutation is that the first unlucky mutant usually dies or is made sterile and is thus not able to pass on the pseudogene. A tragic fate for her or him, but lucky for the rest of us.

Some scientists refer to pseudogenes as dead rather than broken, because nature has managed to "resurrect" some of them to serve new functions. This reminds me of something a friend of mine once did when his refrigerator broke. Instead of undertaking the hassle of hauling it to a junkyard, he turned it into an armoire for his bedroom. He didn't *buy* a refrigerator intending to turn it into an armoire; he just repurposed a broken fridge as one because it was much easier than getting rid of it. He resurrected his dead refrigerator for a brand-new purpose. It was a nifty trick—but as far as we know, resurrected genes are as rare as armoire-refrigerators.

## Alligators in the Gene Pool

As we've just seen, the process of copying DNA is not perfect. The machinery that our bodies have developed for this purpose occasionally makes mistakes, and these mistakes can cause problems. But those kinds of mutations are sporadic—freak accidents, like the sudden death of the *GULO* gene in one primate's genome, which just so happened to spread to an entire population of organisms. Like the mutations themselves, the diseases that sometimes result from these errors, such as scurvy, are sporadic. But there is an entire class of genetic diseases that are more insidious than these, precisely because the mutations that cause them weren't fixed by the accident of genetic drift. They were actually *favored* by natural selection.

Humans have a whole host of persistent genetic diseases that have been with our species for generations, millennia, even *millions* of years. Each of them has an interesting story, and collectively they can teach us some valuable lessons about the sloppy and sometimes cruel process of evolution.

Probably the most well-known and widespread example of a genetic disease that has frustrated humans for ages is sickle cell disease, or SCD. Three hundred thousand babies are born with this condition every *year.* In 2013 alone, at least 176,000 people died from it. The disease is caused by a mutation in one of the genes for hemoglobin, the protein that carries oxygen around in the bloodstream and delivers it to all cells.

Normally, red blood cells are packed with hemoglobin and take on a shape that facilitates both maximum oxygen delivery and optimal folding so that the cells can squeeze through tiny blood vessels called capillaries. The mutant versions of hemoglobin found in SCD patients, however, do not pack together as tightly, resulting in poorly shaped red blood cells. These malformed cells do not deliver oxygen as efficiently and, worse, cannot fold and squeeze through small vessels. They tend to get stuck in tight spaces, creating a sort of sanguinary traffic jam that leads to an intensely painful and sometimes life-threatening sickle cell crisis as the tissue downstream of the traffic jam becomes starved for oxygen. In the developed world, the danger of sickle cell crises can usually be managed with close monitoring and modern medicine. In underdeveloped regions of Africa, Latin America, India, Arabia, Southeast Asia, and Oceania, it is often fatal.

The strangest aspect of SCD is that it is caused by a single-point mutation, a simple switch of one DNA letter for another (though there are many possible point mutations that can cause the disease, with different mutations common to different geographical ethnic groups). This is truly strange because a point mutation with such a drastic negative effect on survival is usually eliminated from the population rather quickly. Research in population genetics has shown convincingly that a mutation that causes even a slight disadvantage will be eliminated from a population in a matter of a few generations, not thousands of years. To be sure, genetic diseases caused by the interaction of multiple genes or those that give only a slight predisposition toward illness are sometimes difficult for natural selection to sort out. But SCD should be

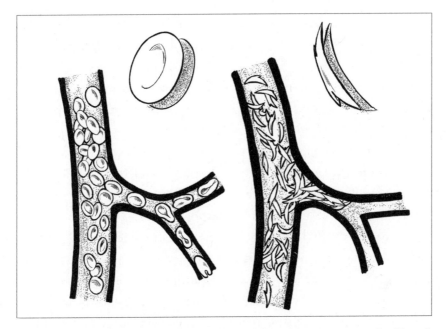

The shape of normal red blood cells (left) and those showing signs of sickle cell disease (right). Whereas normal red blood cells easily fold in half in order to squeeze through tight capillary vessels, sickle-shaped cells are much less flexible and often get stuck in tight spots.

easy. It's a single mutated gene with a disastrous effect. There is simply no way that it should have persisted for very long.

Yet the mutant coding that causes SCD is hundreds of thousands of years old, and it has appeared and spread — spread! — in many different ethnic groups. How could a mutation that causes a horrific, debilitating disease, one that can easily be fatal without modern medical intervention, not only pop up multiple times and in multiple places throughout human history but also appear to often have been *favored* by natural selection? Further, how could it have spread so relatively far in the populations it affects?

The answer is surprisingly simple. Like many genetic diseases, sickle cell disease is recessive. This means that you need to inherit two copies of the mutant allele, one from each parent, in order to develop the disease. If you inherit only one copy, you will not be affected in any noticeable way—although you will be a carrier, able to pass the gene to your children, and they may develop the disease if the other parent also gives them a bad copy. If two carriers of SCD procreate, approximately one-fourth of their children will develop the disease, even though both parents appear to be healthy. For this reason, recessive traits sometimes give the appearance of skipping generations. Nonetheless, with SCD being so deadly, the early deaths of the sufferers of the disease should have eventually eliminated it from the population.

The reason the SCD mutations were *not* eliminated is that the carriers of SCD—the individuals who have only one copy of the code and are thus not afflicted with the symptoms—are more resistant to malaria than noncarriers. Malaria, like SCD, is a disease affecting red blood cells. However, malaria is caused by a parasite that is passed to humans through mosquito bites. It turns out that people with just one copy of the mutant SCD allele *do* have a slight difference in the shape of their red blood cells. It's not enough to cause the sickle cell disease, but it is enough to make the cells inhospitable to the parasite that causes malaria.

Sickle cell disease is often discussed in introductory biology courses as an example of something called heterozygote advantage—*heterozygote* being the term for someone who has two different copies of a certain type of allele. A carrier of SCD is a heterozygote for the affected gene because she has one copy of the mutant allele and one normal copy. To see why being a carrier could be an advantage, first consider the fact that, if you get two copies of the mutant SCD allele, you are in serious trouble. However, if you get just one copy, you are better off than those who have no copies, because you're both SCD-symptom-free *and* you have a lower chance of contracting malaria. In areas of the world where malaria is and has been a real problem, the mutant SCD gene is being

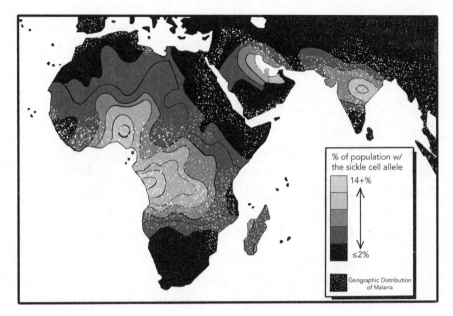

A map comparing global distribution of the genes for sickle cell disease with the range of the Plasmodium parasite that causes malaria. Because sickle cell disease confers resistance to malaria, their geographical reaches are markedly overlapping.

pulled in two directions by natural selection. On the one hand, sickle cell disease can be deadly; on the other hand, malaria can also be deadly. Evolution had to weigh one threat against the other, and the result is a compromise in which the mutant alleles that cause SCD but protect against malaria are present in up to 20 percent of the population in the most malaria-infested regions of Central Africa.

Quite predictably, sickle cell disease is not distributed evenly across human populations. After all, an SCD mutation that appeared in an individual living in a region relatively free of mosquitoes and malaria —say, Northern Europe—would confer no advantage on that individual. The allele would simply be a disease-causing mutation and would thus not persist. For this reason, sickle cell disease is almost unheard of in European populations. In fact, the geographic prevalence of sickle cell disease overlaps amazingly well with the geographic prevalence of malaria.

There's an interesting final twist in the story of sickle cell disease. Researchers understood the push-pull of evolutionary pressure on the mutant code, but they initially could not understand why SCD persisted, because sickle cell disease is so much more deadly than malaria. They built computer models that predicted the opposite: SCD should have died out. But they overlooked a crucial factor stemming from the fact that many preagricultural human societies engaged in polygamy.

In most polygamous societies, a minority of men have multiple wives, meaning that males directly compete with one another for the privilege of reproducing with as many females as possible, with most males not reproducing at all. Among males, competition is quite fierce, and there is a direct and strong relationship between the number of offspring one leaves and one's overall health, vitality, and virility. In this scenario, the push-pull of malaria versus sickle cell disease would likely doom males that had either two copies or no copies of the SCD alleles — these males would be either afflicted by sickle cell disease or vulnerable to the ravages of malaria. Most of the dominant and prolific males would thus be carriers. These alpha males (to borrow a loaded term) would reproduce with a large number of females and produce an even larger number of offspring. Many of those offspring wouldn't make it to adulthood because, in addition to having the normal afflictions, infections, and deficiencies of premodern life, many of them would also have to contend with either malaria or sickle cell disease. That would have been perfectly okay, however, because the SCD carrier and his harem would constantly produce more babies.

Compared to binary marriage, polygamy substantially intensifies the selective pressures of health and survival because of the direct competition of males with one another. The heterozygote advantage is even stronger because a male must be in tiptop shape in order to fend off other males and earn a harem. Any predilection toward SCD or sensitivity to malaria would be a weakness he could ill afford. Although polygamy was never a universal human practice, it was common enough in certain places at certain times to facilitate the proliferation of the gene mutations that cause sickle cell disease. Many people whose ancestry traces to malaria-infested tropical regions still suffer from this genetic malady.

Other single-gene genetic disorders include cystic fibrosis, various forms of hemophilia, Tay-Sachs disease, phenylketonuria, Duchenne muscular dystrophy, and hundreds more. These genes are recessive, like the gene for SCD, so you must inherit the mutation from both parents in order to be afflicted. This makes each of these diseases fairly rare. However, collectively, genetic diseases are not uncommon; some estimates suggest that genetic disease affects around 5 percent of the human population. While not all of them will be lethal or even debilitating, we are still talking about hundreds of millions of people walking around our planet with errors in their genetic code. Most of those errors occurred generations ago, and many of the people who carry those errors don't even know it because they are heterozygote carriers. Those who do suffer are the result of an incredibly unlucky pairing of two unknowing carriers.

A few genetic diseases are caused by a dominant mutation rather than a recessive one. This means that, instead of having to inherit a bad gene from each parent, just one from either parent is enough to cause the disease. These are much rarer for the obvious reason that they aren't ever hidden; the selective pressure against a genetically dominant disease mutation is usually swift and unforgiving. However, a few of these mutations have persisted across generations, as have the genetic diseases they cause: for example, Marfan syndrome, familial hypercholesterolemia, neurofibromatosis type 1, and achondroplasia, the most common form of dwarfism. These conditions are often inherited from a parent, but even when the mutation occurs spontaneously in an individual with no family history of it (which actually happens quite often), the resulting disease will be passed on to 50 percent of that person's offspring. Sadly, therefore, genetic diseases are just as persistent in the lineages of people who developed the mutation sporadically as they are in the lineages of individuals who inherited their condition from one of their parents.

Among the most well-known genetically dominant diseases is Huntington's, an especially cruel disease. Symptoms don't usually appear until the victim is in his or her early forties to late fifties. Following the onset of the disease, patients suffer a slow deterioration of the central nervous system, beginning with muscle weakness and poor coordination and progressing to memory loss, mood and behavior changes, loss of higher cognitive functions, paralysis, vegetative state, and eventually coma and death. The deterioration is excruciatingly slow, taking five to ten years, and as of now there is no cure or even treatment to slow the progression. The patients and their loved ones are powerless, fully aware of the nature of the impending decline.

The cause of Huntington's disease, like all genetic diseases, is a mutation in the genome. However, if a genetic disease persists instead of being eliminated by natural selection, there must be a reason, as we saw with sickle cell disease. This is especially true in the case of a genetically dominant condition like Huntington's, which has no carriers — only victims. Around one in every ten thousand people in Western and Northern Europe has the dominant and lethal Huntington's mutation (with the highest rates in Scandinavia and the British Isles). That may not sound like much, but it adds up to hundreds of thousands of people in these regions alone. Although the percentage of Huntington's mutations in Asian populations is much lower than it is in Europe, the total number of people who are affected by the disease is much higher, given how much more populous Asia is. This begs the question: How can Huntington's disease be so common if it is also so deadly?

The answer is almost as cruel as Huntington's itself. By the time

Huntington's strikes, the sufferer is past prime reproductive age and so may have already passed the disease gene on. Rather than taking the disease allele with them when they die, the victims leave it behind with their offspring as a grim genetic legacy.

The genetics behind Huntington's disease wasn't even discovered until the late nineteenth century. Before then, no one had any idea that it was passed on in such a simple manner. Of course, it seems obvious now, but the nature of Huntington's was partly obscured by the fact that, until the past two or three centuries, most people died before reaching age forty. The disease didn't track as cleanly in a family tree as it does now because people were dying of so many other afflictions and infections before they reached the ripe old age of two score years. In addition, in earlier times, both women and men tended to begin their reproductive lives earlier than people in the developed world do now. When someone did live to be forty years old, he or she was likely to be an aged grandparent, and a disease like Huntington's, with its slow start and nonspecific early symptoms, was mistaken for dementia or just old age.

Because of its late onset, Huntington's can be passed on with natural selection having little to say about it. Selective forces can operate only on inherited traits that directly or indirectly affect reproduction or survival—that is, survival through reproductive age. Beyond that, one's genes have already been passed on to the next generation's gene pool. An affliction like Huntington's doesn't much affect the number of successful offspring one produces, so it is largely in natural selection's blind spot.

Genetic diseases are shockingly common in the human population and are often deadly or debilitating. Most are inherited, and whether they have persisted for generations or are the result of a sporadic mutation, they all come down to errors in our DNA blueprints. Chromosomes break, DNA is mutated, and genes are destroyed. And evolution is sometimes powerless to stop it.

As if that weren't bad enough, there is another onslaught that our genomes must endure: viruses.

## Our Viral Graveyards

Just as it's populated with pointless pseudogenes and harmful disease genes, the human genome also contains the remnants of past viral infections. Strange as this may seem, these viral carcasses are widespread; as a percentage of the total DNA letters in your body, you have more viral DNA than genes.

You have ancient viral DNA in all of your cells thanks to a family of viruses called retroviruses. Of all the kinds of viruses that infect animal cells, retroviruses may be the most nefarious. The life cycle of a retrovirus includes a step in which its genetic material is actually inserted into the genome of the host cell, like a parasite made of pure DNA. Once ensconced in the tangles of genetic material, it waits for the perfect time to strike—and when it does, the results can be catastrophic.

HIV is the best understood retrovirus. When HIV enters a human T cell, the virus consists of little more than a few genes made of RNA (another genetic-code molecule closely related to DNA) and an enzyme called reverse transcriptase, or RT. After the virus unpackages itself to begin the infection, the RT enzyme makes a DNA copy of the viral RNA. This DNA copy then nestles inside the host cell's DNA on some unsuspecting chromosome. Once it is integrated, it can lie in wait indefinitely, perfectly hidden within the endless strands of As, Cs, Gs, and Ts of the host cell. It can pop out and pop back in at will. When it pops out, it engages in an active phase of viral attack. When it pops back in, the virus goes dormant. This is why people with HIV can have occasional bouts of severe illness followed by periods of relative good health.

This is why HIV is still impossible to cure. It lives in the DNA. There is no way to kill the virus without also killing the host cell. Killing all T cells is not an option because then we wouldn't have working immune systems. Instead, recent therapies that have had great success in treating HIV are targeted at simply keeping the virus in the dormant phase for the rest of the patient's life.

Of course, the virus is not passed from parents to children genetically (although cross-infection can occur between mother and child during childbirth or before). The reason it's not inherited is that the virus infects only T cells, which have no role in passing genes from parents to children. Only sperm and eggs do that. However, if a retrovirus *were* to infect one of the cells that give rise to sperm or eggs, a child could literally inherit a viral genome from one of her parents. She would be born with a virus hidden inside the chromosomes found in every single cell of her body, like trillions of tiny Trojan horses waiting to unload their malicious contents on their unsuspecting host. Her parent had the virus only in a sperm- or egg-producing cell. She has it everywhere!

This inherited viral DNA need not produce an active infection in order to be propagated. In fact, it doesn't need to produce any actual viruses at all. The genome of the virus, once inside the core DNA, will be passed on no matter what. For the virus, this is an absolute victory; it doesn't have to do any other work in order to spread.

This is precisely what has taken place countless times in human history, and the resulting viral carcasses are still with us. Thankfully, they're now heavily mutated after all this time, to the point that almost none of them are able to create infections. (Although, as we'll see, even dead viral DNA can and does do harm.)

Around 8 percent of the DNA inside every single cell of your body consists of remnants of past viral infections, nearly a hundred thousand viral carcasses in all. Humans share some of these carcasses with cousins as distant as birds and reptiles, meaning that the viral infections that originally created them took place many hundreds of millions of years ago and these viral genomes have been passed along, silently and pointlessly, ever since.

Truly, most of these viral corpses serve no function whatsoever, even as the body dutifully copies each one of them hundreds of millions of times a day. The good news is that all or nearly all of our parasitic viral genomes have quieted down to a truly carcass-like state, never doing any work such as, um, releasing active viruses into our cells. (Here's a premise for a sci-fi thriller: An evil genius discovers how to turn on the ancient, dormant viruses skulking in our DNA. Our bodies would destroy themselves from within, probably in a hurry.)

Yet while they're mostly dormant, these genetically inherited viruses do have a blood-soaked past—one that occasionally seeps into the present. They have surely killed innumerable individuals over the years due to their tendency to disrupt other genes. Retroviral genomes can jump around and insert themselves randomly into chromosomes; like a bull in a china shop, they cause all kinds of damage because, even though they can't make viruses anymore, they retain their pop-in, pop-out abilities, and if they pop into an important gene, they can do great harm. As if this weren't odd enough, it turns out that pieces of *our own* DNA can also jump around through the genome.

## Jumpy DNA

I have saved perhaps the most perplexing, and certainly the most abundant, type of pointless DNA for last. Lurking in our genomes are regions of highly repetitive DNA called transposable elements (TEs). TEs are not genes; they are pieces of chromosomes that can actually get up and move around, changing position during cell division, not unlike the retroviral genomes discussed above.

If that sounds weird to you now, imagine how absurd it seemed when first proposed by Barbara McClintock in 1953. Her theory was the only explanation she could find for the bizarre genetic phenomenon of haphazardly inherited colored stripes on corn leaves. The scientific community was completely incredulous and dismissed her ideas with barely a second thought, but she worked tirelessly to refine and advance her theory anyway, putting it to the test in hundreds of painstaking experiments with corn plants. More than twenty years after she first proposed the existence of TEs, they were discovered in bacteria, and this time by more "traditional" research groups (I use these quotes cynically, to mean "led by men"). This forced the scientific community to take another look at McClintock's work and admit that she was right. In 1983, she was awarded the scientific community's highest honor— the Nobel Prize.

One particular TE, called *Alu,* offers a good example of how these curious elements of our genomes—these "jumpy" bits of DNA—have come to be what they are. We know the most about the *Alu* element because it is the most abundant TE in humans and other primates. There are *one million* copies of it in the human genome. These copies have been dispersed all over the place, on every chromosome, within genes, between genes—everywhere. The story of how they ended up in the human genome is both incredible and completely improbable.

Once upon a time, in the genome of a creature that inhabited the earth over a hundred million years ago, a gene called *7SL* did something strange. Every living cell in every organism today, from bacteria to fungi to humans, has a version of *7SL,* which helps build proteins. However, in a sperm or egg cell of some ancient mammal, a molecular mistake was made. Two *7SL* RNA molecules were fused together, head to tail. Coincidentally, a retrovirus infection was ravaging the same cell, and one of the viruses inadvertently grabbed this misshapen double-*7SL* RNA molecule and started making DNA copies of it. These DNA copies then inserted themselves back into the genome of this anonymous mammal's cell, creating multiple copies of *7SL:* one normal version (which we still have), and many of the fused copies. Not knowing it was anything unusual, the cell transcribed the fused *7SL* genes into

RNA as if they were normal genes. The retrovirus again took the RNA products and made DNA copies from them. Some of those copies inserted themselves into the genome, and the cycle continued over and over again, amplifying exponentially each step of the way. We cannot know how many *7SL*-fused elements, which we now call *Alu,* that the cell and virus initially made, but it was certainly thousands.

Through pure chance, the offspring that resulted from that sperm or egg cell became the ancestor of a whole group of mammals called the supraprimates, which include all rodents, rabbits, and primates. We know this because all those animals contain many hundreds of thousands of copies of the bizarre *Alu* element but no other animals do.

You'd be forgiven for thinking that a molecular accident of this magnitude—one resulting in hundreds of thousands of copies of a freakishly deformed gene being scattered throughout an organism's genome—would have had serious, negative ramifications for the animal in which it occurred. The fact that supraprimates are here means, of course, that it didn't, at least not right away. Most of these copies and insertions fell harmlessly into sections of DNA that simply don't matter much, if at all. The *Alu* sequences spread from this originating organism to its offspring, eventually becoming fixed in that ancient species and all its descendants. *Alu* has since taken on a life of its own, continuing to copy, spread, mutate, insert itself, reinsert itself, and just generally bumble its way through the genome. Most of that bumbling is harmless, but occasionally it can wreak havoc.

In fact, we don't have to look deep into our evolutionary past to identify the harm that one million random insertions can do when they rip through the genome. To this day, genetic damage caused by rogue *Alu* insertions makes humans more susceptible to a variety of diseases. For instance, *Alu* and other TEs created the "broken" gene alleles responsible for hemophilia A, hemophilia B, familial hypercholesterolemia, severe combined immunodeficiency, porphyria, and Duchenne muscular dystrophy. *Alu* went crashing into these important genes and either completely destroyed or severely disabled them. Disruptions by *Alu* or other TEs have also created genetic susceptibilities to type 2 diabetes, neurofibromatosis, Alzheimer's disease, and cancer of the breast, colon, lung, and bone. These are genetic *susceptibilities,* meaning that the gene was weakened rather than completely destroyed. Nonetheless, this genetic damage has undoubtedly killed millions of humans just in the past few generations.

The existence of transposable elements may seem like a real failure of evolution; after all, shouldn't natural selection eliminate harmful genetic material like this? But the thing to keep in mind is that evolution often operates not just at the level of the individual but also at the level of the gene or even a small stretch of noncoding DNA. Yes, random mutations may be disadvantageous for an individual and yet still persist through the power of their own self-copying. This was Richard Dawkins's big insight described in his book *The Selfish Gene.* If a small bit of DNA such as *Alu* can act to promote its own duplication and proliferation, it will be favored by natural selection regardless of whether it harms the animal host unless it hurts the animal host so much that the host dies before it can reproduce, which of course does sometimes happen. But in *Alu*'s case, this little piece of genetic code has proven itself so proficient at reproducing that it can more than withstand the occasional host death due to the element's overdoing it.

If we add up all the various *Alu* sequences — more than a million copies spread throughout your DNA — this one particular molecular parasite makes up more than 10 percent of the total human genome. And that's just *Alu*. If you add up all the TE insertions, it comes to about 45 percent of the human genome. Nearly half of human DNA is made of autonomously replicating, highly repetitive, dangerously jumping, pure genetic nonsense that the body dutifully copies and maintains in each one of its billions of cells.

## Coda: Color Me Lucky

As you'll see again and again in this book, certain flaws are built into nature. They are not bugs in the system; they are features (so to speak). Thus, the fact that each of us is carrying and reproducing one million useless *Alu* sequences in our DNA is certainly an oddity, but at this point, it is also an inherent quality of our bodies. And like some other flukes that have become features, *Alu* has resulted in some very rare and totally unexpected benefits.

The way that *Alu* has helped us is in its tendency to create mutations, those almost always damaging but occasionally helpful changes to DNA. By jumping around the genome, *Alu* raises the mutation rate of the organisms that have it, and it can even occasionally cause chromosomes to break in half. While that sounds awful, because mutations and damage to chromosomes are almost always bad for the individuals in which they occur, it can actually be beneficial over the long term. This is because a lineage of animals with high mutation rates will be more adaptable and thus more genetically malleable over long periods (assuming they don't go extinct because of all those mutations).

Although it's cold comfort to the individuals that suffer and die from harmful mutations caused by *Alu,* the rare emergence of a helpful mutation can change the course of evolution in dramatic ways. We have to take a very long view to appreciate this, but rare beneficial mutations provide the raw material with which natural selection produces new adaptations. The most famous example of this is the mutation that led to our species' excellent color vision.

Around thirty million years ago, a random *Alu* insertion occurred in an ancestor of all Old World monkeys and apes (including humans) that would allow a subsequent improvement in the ability to see a rich variety of colors. In our retinas are structures called cones that specialize in the detection of specific wavelengths of light—in other words, colors. These cones have proteins called opsins that respond to different colors, and, prior to thirty million years ago, our ancestors had two versions of the opsin proteins, each responding to a different color. Then a happy genetic accident occurred due to something we call gene duplication.

Simply put, an *Alu* element, doing its normal business of crashing around the genome, popped into a chromosome very near one of the opsin genes. It copied itself and popped out, but it had inadvertently copied the opsin gene, complete and intact, and it brought the copy along for the ride. When this newly copied *Alu* element popped back in the genome somewhere else, it brought the copy of the opsin gene with it. And voilà, this lucky monkey went from having two versions of the opsin gene to having three. This is called gene duplication.

Gene duplication is normal behavior for *Alu*—that's why we have so many duplicates of it—but it's nearly miraculous that the tagalong opsin gene was perfectly copied and reinserted in the process. At first, the extra gene would have been identical to the one it was copied from. However, once this species had three opsins instead of two, the three genes were free to mutate and evolve separately. Following some refinement through mutation and natural selection, these ancient monkeys had three types of color-sensing cones in their retinas instead of just the original two. All descendants of these monkeys, including us, have three different types of cones, a trait called trichromacy.

Trichromacy is a vaunted attribute among animals because having three cones instead of just two allows the retina to see a broader spectrum of colors. Apes and Old World monkeys can see and appreciate a much richer color palette than can dogs, cats, and our more distant cousins the New World monkeys. This enhanced color detection served our ancestors very well in their rainforest habitats. Because their *GULO* gene had been broken millions of years before, finding fruit was very important to these monkeys and apes, and having much enhanced color vision is a big help in the hunt for ripe fruit in a dense forest. And here's the kicker: We owe our superior eyesight to a mutation caused by a roaming *Alu* element.

The duplication of the opsin gene and the resulting trichromacy occurred through a series of extremely improbable events, but that's evolution for you. Crazy stuff happens. Most of it is bad—but when it's good, it's *really* good.

# 4

# *Homo sterilis*

*Why humans, unlike other animals, can't easily tell when females are ovulating and thus when the time is right to conceive a child; why human sperm cells cannot turn left; why, of all the primates, humans have the lowest fertility rate and the highest mortality rate for infants and mothers; why our enormous skulls force us to be born way before we are ready; and more*

One of the preconditions of evolution, perhaps the most important, is that a species must be able to reproduce—a lot.

This is because life in nature is a constant struggle. In all species except us (thanks to modern medicine), most individuals that are born will not live to sire offspring of their own. This was one of Darwin's key insights. He noticed that all organisms seem to reproduce constantly and in great numbers and yet their populations remained pretty much the same size. This meant that life was a challenge that most individuals failed.

The only way a species has any chance to survive and compete is by making a lot of babies. Some make fewer babies than others but care for them better, while others make tons of offspring but don't care for them at all. But for all species, prolific reproduction is a key goal of an individual's life, if not *the* key goal. We all have an inborn drive to make more of ourselves. It's the only way a species survives.

Of course, living creatures, even humans, don't really think about reproduction in those goal-driven terms. We want our offspring to survive because of a deep-seated, instinctive parental urge, not because of a conscious desire to preserve our genes for posterity. But the fact remains: we are hardwired to want to pass on our genes.

There is only one way that living things can secure their genetic legacies. They must be sure that at least one or two of their offspring will survive, thrive, and have offspring of their own. It is practically guaranteed that many offspring will die; if a predator or a rival doesn't get them, an infectious disease will. The intensity of natural selection has thus given all animals an extreme drive to reproduce.

Given that humans have successfully outcompeted every other species on the planet, you might think that we've mastered this whole reproduction thing. But in fact, human reproduction is inefficient. *Extremely* inefficient. We are some of the most inefficient reproducers in the animal world because we have errors and flaws throughout almost the entire reproductive process, from the production of sperm and eggs to the survival of our children. I describe this as inefficient because a breeding pair of two mammals ought to be able to produce way more offspring than humans do, and most other mammals do a better job at it. If you started with two fertile cats, in a year or two, you could have hundreds. If you start with two humans, after a couple of years, you *might* have one more. Yes, humans take longer to gestate and mature, but that's not the only limitation in play, as we'll soon see.

Human reproductive inefficiency is far out of step with the reproductive abilities of other mammals, including our closest relatives. Strangely, there are very few explanations as to why this is. For some aspects of our reproductive difficulties, we understand the cause, but for most, we don't. Human beings are riddled with fertility problems.

It's hard to believe that we're such inefficient baby-makers, given that the world's human population now exceeds seven billion. But in a way, our shortcomings in this department make our terrific evolutionary success all the more impressive.

## Infertile Myrtle

It may be tempting to blame our reproductive inefficiency on one big problem; for instance, on our huge brains, which require huge skulls, which makes childbirth perilous for mother and infant alike. But it is not that simple. The entire reproductive process—from the production of sperm and eggs to the survival of infants—is plagued with problems that highlight a wide range of design defects in the human reproductive system. In practically every part of that system, human beings have more faulty biology than any other mammal we know of. Something is seriously wrong with us in this regard.

You could argue that these inefficiencies are somehow adaptive; perhaps they serve a purpose, such as controlling population growth. Although I will discuss this possibility shortly, it is worth noting now that, if this were indeed the case, it would be a pretty dismal compromise. Other species have achieved the same end through much more elegant means. For instance, helper wolves forgo their own reproduction and instead care for their kin, but there is nothing amiss in their *bodies;* their reproductive anatomy is just fine. The social structure is such that some wolves choose to remain celibate—a choice that can be reversed if the alpha wolf in their group dies or is vanquished.

Not so with humans. For many individuals, infertility is not a choice, and often it is not reversible without the aid of medical advances, most of them very recent. Besides being biologically incongruent, moreover, any comparison with helper wolves—or worker bees, or drone ants, or other creatures that sacrifice their own reproductive chances for the good of their group—is downright cruel to anyone who has experienced the frustration and pain of infertility. And these people number

in the *millions*. A staggering percentage of humans experience reproductive difficulty, either for extended periods or permanently.

This reality is even more staggering when we consider that infertility
can run in families—take a moment to absorb the bitter irony—and
that there is usually no outward or inward symptom. Worker bees and
helper wolves *know* their roles and the lack of reproduction that goes
with it, and so do their peers. Humans, by contrast, almost never have
any idea that they have fertility issues until they try to conceive.

We all know someone who has had trouble reproducing for one reason or another. Estimates vary based on geographic location and precisely how the term *infertility* is defined, but most studies report that
somewhere between 7 to 12 percent of couples trying to conceive have
faced persistent difficulties. Fertility problems are equally common in
women and men, and in around 25 percent of cases, *both* partners find
that they have reproductive problems.

As many sufferers know, fertility problems have a unique and dis-
proportionate effect on mental health. There are hundreds of diseases
and afflictions that are far more physically debilitating but do not cause
as much emotional anguish. Even the thought of not being able to
have children strikes many people someplace deep in their souls. Most
humans are driven to reproduce, and when they fail, it cuts them to
the marrow, crushing their spirits and self-confidence—this despite the
fact that not even the most heartless among us could possibly assign
victims of infertility blame for their condition.

For all the stigma and shame attached to infertility, we've all been
infertile at one point in our lives. I'm talking, of course, about the infertility that humans experience before they reach sexual maturity—a
fallow period that you might not think of as infertility, per se, but that
nevertheless has a very similar effect as adult fertility when it comes to
the reproduction of the species.

First of all, humans mature rather late compared with most other
mammals, even compared with our closest relatives. Humans mature,

on average, two to three years later than chimps and four to five years later than bonobos and gorillas. Of course, there are good reasons for this. Given the large size of a human baby's head, it is important that the pelvis of the female be large enough to accommodate it during birth. If a woman is of small stature, her chance of dying in childbirth is extremely high—even higher than it already is. (More on that later.) This doesn't explain the lateness of male puberty, which is later even than female puberty, but this is largely without effect in terms of the reproductive capacity of the species. Males and their sperm are never the limiting factor for reproduction of a species even if many or most males happened to be infertile.

The lateness of human female puberty compared with that of other primates results in reduced reproductive efficiency for the species. This is because delayed fertility brings a greater chance that a female will simply not live long enough to reproduce. Remember that as humans were living and dying in the Pleistocene epoch, and even through the Stone Age and the early modern period, life in the wild involved a whole lot of sudden and tragic deaths. This means, in effect, that every year a female was not reproducing increased her chance of dying without leaving any offspring. While this is not such a big deal today, it would have presented a significant challenge to our species for much of its existence. Until the advent of modern medicine, human mortality rates were quite high throughout their lifespans, not just at the far end of it as they are now. For most of our history, many humans died young —and therefore died childless.

The age of sexual maturity is thus the first limiting factor in reproductive capacity, a phenomenon that is true in all species, not just humans. For example, when officials are considering which threatened or endangered species are most in need of regulatory protections, the age of reproductive maturity is a crucial factor. Bluefin tuna, for example, are often cited as an example of a fish species that is in need of protection, not just because of decades of overfishing but because the females do not reach sexual maturity until twenty years of age. This means that a population devastated by overfishing will be very slow to rebound.

But even beyond the extended prereproductive years of the human lifespan, even after humans reach sexual maturity, they often have trouble producing high-quality sperm or eggs, the all-important vehicles of genetic transmission.

Let's start with men. A 2002 study by the CDC found that around 7.5 percent of men under the age of forty-five had visited fertility doctors. While the majority of those were diagnosed as "normal," meaning nothing obviously wrong was found, around 20 percent of them had substandard sperm or semen, making reproduction by the old-fashioned route highly improbable or impossible.

Normally, sperm are amazing little swimmers. Although they're among the tiniest of human cells, they are easily the fastest. After being propelled into a vagina, a sperm must swim around 17.5 centimeters to reach the egg. The sperm cell itself is only around .0055 centimeters (55 micrometers); this is quite a long way, more than three thousand times the length of its body. That's like a human running over thirty kilometers, almost nineteen miles. Even more impressive, sperm swim at around 1.4 millimeters per second, which would be like a human running twenty-five miles per hour—a speed that would allow an individual to cover those nineteen miles in around forty-five minutes. When you consider that Usain Bolt, the world's fastest runner, can reach speeds like this for only a few hundred meters at a time, that feat becomes all the more impressive.

However, men's sperm take much longer than forty-five minutes to travel the distance from the vagina to the fallopian tube. This is because they waste a lot of time swimming around in random directions.

Human sperm cells, you see, can't turn left. This is due to the corkscrew nature of their propulsion system; rather than snapping their whiplike tails back and forth and side to side, sperm cells rotate their tails around in a corkscrew motion, like the way you would move your

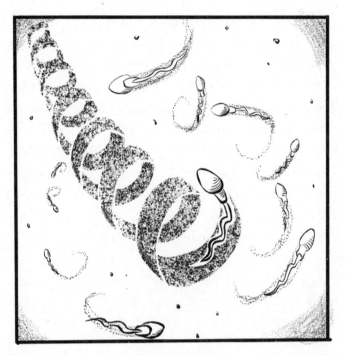

Sperm cells use corkscrew-like locomotion and thus tend to swim in right-hand circles with a random overall trajectory. For this reason, sperm cells actually traverse an extremely long path just to travel the very short distance to the fallopian tubes.

index finger to draw a circle in the air. Because most sperm whip their tails in a right-handed spiral, the rotation pushes them forward and toward the right and they end up swimming in ever-widening circles. This means that it can take as long as three days to reach the egg waiting in the fallopian tube to be fertilized. Very few of the original number get anywhere near their goal. This is one reason why human males produce sperm in such large numbers. You need about two hundred million of them to start with in order to get just one to its destination.

Low sperm count is the most common fertility problem in men. Somewhere between 1 and 2 percent of men suffer from it. These men produce "only" one hundred million (or fewer) sperm per ejaculation. Because the volume of ejaculate varies widely, sperm count is generally measured as sperm per milliliter. While medical professionals do not always agree on what constitutes a healthy sperm count, the average is around twenty-five million per milliliter. Below fifteen million is considered low, and below five million is considered very low; conception by standard means is very unlikely for men who have sperm counts in this range. In some cases, the problem is hormonal or anatomical, and sometimes a combination of medicine and lifestyle or diet changes can restore a healthy sperm count. In the vast majority, however, the best that can be done is a modest increase in sperm count through diet and lifestyle changes.

Low numbers are not the only problem that men encounter when it comes to their sperm. Sperm can also have low motility (meaning that they're slow), poor morphology (meaning that they're misshapen), or low vitality (meaning they're mostly dead). If the semen's pH levels, viscosity, or liquefaction times are abnormal, it can make conception harder as well. In short, a lot can go wrong.

Females have a similar set of problems pertaining to the production and release of eggs. The female's reproductive system is far more elaborate than the male's and is thus more vulnerable to complications. While the majority of these occur in the uterus, thereby impairing a woman's ability to sustain a pregnancy, some women have difficulty releasing healthy eggs in the first place.

About 25 percent of female fertility problems can be traced to a failure to ovulate a healthy egg. In most cases, we just don't know what causes the problem, although a few genetic and hormonal syndromes have been identified as culprits. Fortunately, modern science has had reasonable success in coaxing the female reproductive cycle back on track. With carefully timed hormone injections, many women can be induced to ovulate even when their own hormones have failed them. These treatments work so well that more than one egg is released at a time. This has led to a marked increase in the rates of fraternal twins in Europe and North America.

Even when each potential parent produces and releases healthy sperm or egg cells, there is no guarantee they'll be able to create a pregnancy. First of all, insemination has to be timed very carefully with ovulation or it won't be successful. In a typical twenty-eight-day menstrual cycle, the fertile period is just three days long in the very best of scenarios, with twenty-four to thirty-six hours being the more common window of opportunity. This means that even perfectly fertile couples usually have to try for months before the female conceives.

The biggest obstacle to getting the timing right is the completely concealed nature of ovulation in humans. Neither the male nor the female knows for sure when it happens. This is in stark contrast to basically all other female mammals, including the other female apes, who conspicuously advertise when they are at the fertile point of their estrous cycle. To be sure, other animals have plenty of sex outside the fertile period, underscoring the many nonprocreative functions of sex, such as strengthening a pair bond. However, when the goal *is* to have offspring, it sure is convenient to be clear about the best time for conception.

Why is concealed ovulation peculiar to *Homo sapiens*? There may be adaptive purposes for the concealment; if a man cannot be sure when a woman is ovulating, he cannot be certain that a child is his own unless he sticks with one female constantly. If ovulation were obvious, an alpha male could simply have sex with every ovulating female, spreading his genes widely but not sticking around to invest in the offspring. Thus, concealed ovulation has led humans to form more long-lasting pair bonds and enhanced paternal investment in the offspring. But here, too, a feature of our bodies is also a bug—concealed ovulation adds tremendously to the inefficiency of human reproduction. Other animals know exactly when the fertile period of the estrous cycle is. Humans have to guess.

Most other mammals are so successful at conceiving that females automatically go into the pregnancy cycle right after having sex, even if they're not actually pregnant. For example, in rabbits and mice, if a vasectomized male copulates with a female, her uterus will prepare to nurture a developing pup for many days afterward, a condition called pseudopregnancy. So successful is sexual reproduction in these animals that any time a female has sex while she's in heat, her body just *assumes* conception has occurred.

If human females actually became pregnant every time they had sex during the fertile period, humans would be reproducing like, well, rabbits. But even when the egg and sperm are healthy, and the sperm finds the egg, and fertilization occurs, it is still no sure bet that a viable pregnancy will result. In fact, many of the most error-prone steps lie beyond the moment of conception.

According to the American College of Obstetrics and Gynecology, between 10 and 25 percent of all recognized pregnancies end in spontaneous abortions (miscarriages) within the first trimester (thirteen weeks). That is probably a severe underestimation because it includes only the *recognized* pregnancies that are lost. From studying fertilization in vitro, we have learned that chromosomal errors and other genetic catastrophes are astoundingly common and these can imperil a potential pregnancy well before it can be recognized. Embryologists estimate that, even with otherwise normal sperm and eggs, 30 to 40 percent of all conception events result in either failure of the embryo to attach to the uterine wall or spontaneous abortion shortly after it has.

Miscarriages beyond the first trimester, while not *as* common as spontaneous abortions during that period, also plague the human reproductive process. Of pregnancies that make it to the thirteenth week, 3 to 4 percent end before the twenty-week mark. Beyond twenty weeks, miscarriages are usually referred to as stillbirths and occur in less than 1 percent of pregnancies. All told, a staggering *one-half* of human zygotes —the one-celled union of sperm and egg—don't make it more than a few days or weeks. Honestly, I sometimes wonder if humans are any more efficient than the mighty oak tree that drops thousands of acorns year after year in the desperate attempt to create one or two saplings.

The most remarkable fact about human fertility is that up to 85 percent of all miscarriages are due to chromosomal abnormalities, meaning that the new embryo has extra, missing, or badly broken chromosomes. Doing the math, this means that when a human sperm and a human egg fuse, the resulting embryo ends up with the proper number of intact chromosomes only around two-thirds of the time. The remaining 15 percent of miscarriages are caused by a variety of congenital conditions, such as spina bifida or hydrocephalus.

Of course, chromosomal problems and other congenital defects become an issue only after a woman has become pregnant in the first place. Sometimes things don't even *get* that far. Even when everything goes right—healthy sperm finds healthy egg in the right place and the right time, and the chromosomes commingle correctly to produce a zygote with nothing extra and nothing missing—pregnancy simply doesn't occur *and we have no idea why not.* This is called failure to implant, and it happens shockingly often. The developing embryo just sort of bounces off the uterine wall and perishes from lack of nourishment.

And even when an embryo *does* implant, it sometimes fails to convince the body not to begin menstruation. This means the embryo has failed at its first challenge: preventing the mother's body from shedding its endometrium (the lining of the uterus and the substrate in which the embryo is living and growing). Embryos have about ten days from the time of implantation to the next scheduled menses, so they typically get right to work secreting a hormone called human chorionic gonadotropin (HCG); this preserves the endometrial lining and thus staves off menstruation, allowing the embryo to continue growing without being thrown out with the proverbial bathwater. Many embryos simply do not secrete enough HCG to prevent the scheduled monthly flow. This means that perfectly healthy, growing embryos are lost in the mother's monthly menstrual blood for no good reason.

While it is not possible to know for sure, conservative estimates for the percentage of perfectly healthy zygotes that fail to implant or fail to prevent menstruation are around 15 percent. (This is in addition to the one-third of zygotes that do not thrive for reasons we *do* know.) Some couples experiencing infertility with no known cause may actually be conceiving zygotes just fine; the embryos may simply be failing to take root in the womb.

These glitches in the reproductive system are especially maddening—and painful—for couples trying to conceive. And they are all bug, no feature; there is no justification for the spontaneous abortion of a perfectly healthy embryo or for the apparent failure of seemingly healthy reproductive organs to establish a pregnancy in the first place.

Considering all the challenges facing couples trying to conceive and maintain a healthy pregnancy, it's kind of amazing that anyone makes it through gestation at all. For those who do, one final danger awaits.

## Death by Birth

Embryos that are lucky enough to have the correct number of chromosomes, implant successfully, and develop properly through the pregnancy must clear one final reproductive hurdle: childbirth. Thankfully, the advances of modern medicine have significantly alleviated the risks involved in this process, but make no mistake: for most of human history, childbirth was an incredibly dangerous endeavor, and lots of children—to say nothing (for now) of mothers—simply didn't survive it.

Global statistics are not kept for the percentage of children who die during childbirth itself. Instead, *infant mortality rate* is generally reported as the percentage of children who do not survive the first year of life—everything between the mother's labor and the child's first birthday.

As of 2014, all but one of the major developed countries had an infant mortality rate below 0.5 percent. The one exception is the United States, which, at 0.58 percent, has a higher infant mortality rate than Cuba, Croatia, Macau, and New Caledonia. (This is due in large part to two particular practices by American doctors: the frequent medical induction of labor, which artificially accelerates the natural process of childbirth, and the overuse of cesarean sections. The reason C-sections are performed so often in the United States? Lawyers. Doctors fear being sued on the off chance that a C-section was needed but wasn't performed. But tragically, these invasive abdominal surgeries often involve many fatal complications of their own.) By contrast, the infant mortality rate in Japan is 0.20 percent, and in Monaco, it's 0.18 percent.

These are relatively low risks, to be sure, but birth is still one of the riskiest moments of our lives. And in regions where medical practice is far from modern, there is still a high rate of infant mortality—a testament to how far from perfect the human reproductive system is. In Afghanistan, for instance, the United Nations estimates the current infant mortality rate at 11.5 percent. In Mali, it is 10.2 percent.

For readers in the developed world, it is astounding to contemplate that in those two countries, one in ten babies does not survive the first year of life. Three dozen more countries, all of them in Africa or South Asia, have an infant mortality rate above 5 percent.

If we look back in time, even within the wealthiest nations, we see a much higher infant mortality rate than we have now. In the United States in 1955, for instance, more than 3 percent of babies did not make it to their first birthday. That rate is six times higher than it is today. In poor countries, the situation was even worse in 1955 than it is now. There were dozens of countries whose 1955 infant mortality rate was higher than 15 percent, and several whose rates were above 20 percent!

My mother had five children, the first in the mid-1960s. Had she lived in Nepal or Yemen a decade earlier, it is unlikely that all her children would have survived. (This is especially unsettling to me, since I am the fifth.) This is all the more disturbing considering that this high level of mortality occurred within living memory, not way back in the Stone Age. How much worse would things have been in the prehistoric period?

This sad state of affairs is in no way representative of other primates or any other mammals. While chromosomal errors and failures to implant are probably just as high among our ape relatives, miscarriages, stillbirths, and infant deaths during delivery are quite rare in other animals, especially primates. One-year infant mortality for wild animals is difficult to measure with certainty, but the best estimates for the other apes is 1 to 2 percent, making their birthing process several times more dangerous than that of humans in the modern United States but several times *less* dangerous than that of people in Mali or Afghanistan or the pre-1950 United States. Keep in mind, too, that we are talking about apes in the wild; animals born in their natural habitat usually fare much better than those born in captivity.

In other words, ultrasounds, fetal monitoring, antibiotics, incubators, respirators, and, of course, expert physicians and midwives have all worked together to bring the human infant mortality rate down to what it is for most other species naturally.

Part of why humans are so out of step with other mammals when it comes to childbirth is because human infants are simply born too early. This is due to our species' massive craniums and the females' relatively narrow hips. Human gestation time is similar to that of chimps and gorillas, even though humans' much larger brains require more time and cognitive development in order to reach their full potential. However, the size of the female pelvis limits how large the fetus's head can grow while still in utero. If it grows too large, there is no way to get it out, and both baby *and* mother can perish. The compromise is that fetal gestation is cut short, and human babies are born way before they are ready.

We are basically all born premature. Premature, and completely

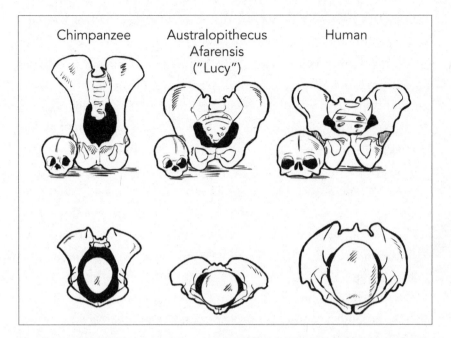

The relative sizes of female pelvises and infant heads in (from left to right) chimpanzees, Australopithecus afarensis (of "Lucy" fame), and modern humans. Human infants' large craniums barely fit through their mothers' birth canals—one of the main reasons why infant and maternal mortality is high in humans but rare in other apes.

helpless. The only thing human infants can do for themselves is suckle, and around 5 percent can't even do that. This, yet again, is not the case for most other mammals (other than marsupials, but they complete their development in a pouch, which is cheating). Baby mammals such as cows, giraffes, and horses hit the ground running—literally. Once they pop out and shake off, they amble around almost immediately. Dolphins and whales are born underwater and, without a moment's hesitation, swim to the surface to take their first breath with little or no struggle. Humans, however, need more than a year before they can get around on their own, and in the meantime, they are vulnerable to any number of threats.

Human infants are so helpless that it almost seems like there must be a reason for their plight—a reason that, perhaps, could also help to explain why we as a species are so bad at making babies in general. Indeed, our myriad problems with fertility represent such a striking contrast with other mammals' reproduction that some biologists have wondered if it may actually be an adaptive response to how helpless humans are as infants.

These scientists reason that a reproductive slowdown was required so parents could give children the time and care they needed before the adults reproduced again. In this view, our reproductive problems are not a curse but a blessing. They have the effect of making pregnancies less frequent than they would be otherwise, which in turn means that each child humans *do* manage to create has a better chance of success because he or she will be the sole focus of parental care for a longer time. In other words, our species' poor overall reproductive rate may be nature's way of keeping parental attention on the helpless infant until he or she can stand on his or her own two feet (again, literally).

There is just one problem with this reasoning. If nature wanted humans to space out children, why achieve it through painful and energy-expensive deaths and false starts? Especially when there is a far easier way: the female body could just delay the postpartum return to fertility for a longer period. That's what many species, including our close relatives, do. In gorillas, average birth spacing approaches four years, except when a nursing infant dies, at which point the mother almost immediately goes into estrus. In chimps, the average spacing is more than five years, and in some orangutans, average birth spacing is nearly eight years! Continued childcare, primarily nursing, inhibits the ovulation-menstruation cycle in these apes, leading to sensible spacing intervals. Mothers and fathers are free to provide the care to their infants and juveniles for as long as the offspring need it.

Not so with our species. Humans keep pushing them out and hoping for the best. Because all our closest relatives have longer postpartum delays in fertility, it is likely that our common ancestors did also. *We* are the outliers, in other words. This means that, through the evolutionary history of humans, we have seen a *decrease* in fertility spacing in females. This doesn't alleviate the problem of helpless infancy; it *compounds* it by throwing additional babies into the arms of parents still struggling to wean previous ones.

The leading explanation for human females' quick return to fertility is that, as human tribes grew bigger, they shifted toward communal parenting. When children were raised cooperatively in large extended families, the parenting burden was shared, and females didn't have to delay the next pregnancy. Furthermore, hunting and gathering became more efficient due to the increasing intelligence, communication, and cooperation of our human ancestors, and this allowed some women to concentrate solely on child-rearing. Not surprisingly, most of the proponents of this theory are men.

Sexist implications alone are not enough for researchers to discard a scientific hypothesis—but there are other reasons to reject this one. For instance, I find this explanation to be insufficient because, at best, it explains only the decrease in birth interval. Humans have issues throughout the entire process of reproduction. If, over the past million years, human advances allowed for an uptick in fertility, why would birth timing be the only sign of that increase, while all other aspects of fertility continued to worsen?

In my view, the rapid rate at which human females become fertile again after giving birth is an accidental byproduct of the evolution of concealed ovulation. Continual and hidden ovulation led to more sex, since neither the males nor the females knew when females were fertile, and thus promoted family cohesion and paternal investment. However, more sex also led to more conception. This happy accident worked out because of the climbing rate of infant mortality due to the aforementioned increase in fetal skull size. Since the babies of humans died far more often than the babies of other apes, the higher birthrate compensated for that loss.

No matter how it evolved, reduced birth spacing with high infant mortality is incredibly poor planning by whatever force designed our species' reproductive system. That shouldn't surprise us, however, because evolution doesn't make plans. It's random, sloppy, imprecise—and heartless.

## Deadly Delivery

Of course, human infants aren't the only ones at risk during childbirth; mothers can and do die from it too. Once again, modern medicine has very effectively managed this risk; in the United States in 2008, for instance, there were only 24 maternal deaths per 100,000 live births. (Shockingly, this was up from 20 in 2004 and 9.1 in 1984, due in large part to the aforementioned overuse of C-sections.) However, in developing countries, the figures are much higher. In Somalia in 2010, there were 1,000 maternal deaths per 100,000 live births. That's 1 percent of all births. When you factor in the much higher birthrate in developing countries, the cumulative lifetime risk for a woman to die in childbirth there is about 1 in 16. Most Somalis will lose several women in their lives to childbirth.

There is great debate regarding the maternal death rate in centuries past, not to mention in the classical periods, prehistory, and preagriculture. It seems that 1 to 2 percent is the *lowest* possible estimate when you consider that several countries (including Somalia) currently have that rate now. Thus, in ages past (as in some places today), childbirth was an incredibly dangerous experience. It is no exaggeration to say that, for most of the time our species has existed, being born was the leading cause of death. For women, giving birth was the next biggest threat.

This, too, sets humans apart. Childbirth is safer for primate mothers in the wild than it is for human mothers, and that's without primates having the benefits of medical intervention. Mothers dying in childbirth is unheard of in chimpanzees, bonobos, gorillas, and all of our other primate cousins. This is a purely human peril.

Especially dangerous for mothers are breech births, when the baby comes out feet-first instead of head-first. It is, of course, possible to deliver a baby in the breech position, but it is much more difficult. Without the aid of medical care, mortality for both baby and mother is higher with breech deliveries. (Estimates vary widely, but all agree that the risk of death or harm in a breech delivery is at least three times higher for the mother and five times higher for the baby. This is the heightened risk in today's world, where women have access to postpartum care and modern medicine.) Much of the risk to baby is due to a tenfold higher risk of constriction of the umbilical cord, which deprives the baby of oxygen. This can last many hours, given the prolonged nature of a breech delivery. For this reason, in these cases physicians almost always opt for delivery by cesarean section.

According to legend, the cesarean section was first performed on Julius Caesar's mother when it was discovered that the infant was presenting in the breech position. This legend is now widely believed to be false, but it is true that the cesarean was well known in the ancient world, where a breech birth meant a very good chance of losing both the mother and the baby. Tales of cesarean births of humans or demigods can be found in ancient Indian, Celtic, Chinese, and Roman mythologies. In fact, long before Julius Caesar, there was a Roman law that stated that when a pregnant woman died, a cesarean had to be performed in an attempt to rescue the fetus. (This probably started as a public-health policy and then apparently morphed into a superstition that unborn infants would experience a ghoulish resurrection if they were buried in their mothers' wombs — giving mourning family members an added incentive to carve into the bellies of the dearly departed.)

There's no more compelling evidence for the riskiness of breech delivery in the ancient world than the fact that, unequipped and terrified, people resorted to slicing women open. In the days before hygienic practices and antiseptic operating rooms, this almost always killed the mother, though it may have occasionally saved the child. All this because of shortcomings in our species' gestational design.

If you've witnessed the birth of another mammal species, you'll know that it is usually *not* a dramatic affair. Cows seem to barely notice when they give birth. Gorilla mothers often continue eating or caring for other children during delivery. The difficulties we associate with childbirth are uniquely human, the product of the rapid evolution of a large cranium together with the failure of evolution to keep up with those changes.

Given enough time, natural selection would surely have sorted this out in any number of possible ways. But the chance of a natural adaptive remedy is now virtually nil, as medical intervention has largely solved the problem of childbirth and removed the negative selection of so many women and children dying in childbirth. This is a triumph of human ingenuity over human limitation. Once again, science has provided solutions to a problem caused by nature. But in the process, science has effectively short-circuited evolution, consigning humans to the faulty reproductive systems that nature gave us.

Any discussion of the mortal danger to women during pregnancy and childbirth would be incomplete without mention of ectopic pregnancies. In science, the word *ectopic* is used to indicate that something exists (or an event happens) in a place where it usually doesn't. In the case of ectopic pregnancies, the location is almost always the fallopian tube. When a fertilized egg implants in a fallopian tube instead of the uterus, it is an extremely dangerous situation, and before the age of modern medicine, it usually led to the mother's death.

When an egg is released from the ovary, it travels through one of the fallopian tubes and eventually reaches the uterus. However, unlike sperm cells, an egg has no flagellum — that whiplike tail — to propel it. Also unlike sperm cells, the egg is surrounded by hundreds of follicular cells that form a protective layer called the corona radiata. None of those cells have flagella either, so the result is that the egg and its crew drift down the fallopian tube rather slowly and aimlessly. It's more or less like a bunch of life rafts tied together and floating in a very large sea. It can take the egg a week or more to reach the uterus from the ovary, despite the distance being a mere ten centimeters.

Each sperm cell, by comparison, is jet-propelled by its whiplike tail. Because the egg moves slowly and the sperm move quickly, conception almost always occurs in the fallopian tube; an ovulated egg is still meandering through the tube when a sperm cell rushes to meet it. (In fact, if it hasn't been fertilized, an egg will usually perish before reaching the uterus — that's how slow it's moving.)

Following fertilization, a series of chemical reactions occur in the zygote as it prepares to start development. About thirty-six hours after conception, the zygote begins rapidly and repeatedly dividing in half. The one-celled zygote becomes two cells. Those two divide again to make four. Four become eight, eight become sixteen, and so on until the embryo grows into a hollow sphere of 256 cells, nine or ten days after conception. Only then is the embryo ready to begin tunneling into the uterine wall and sending out signals to the host body to prevent menstruation. Pregnancy begins. As discussed earlier, halting menstruation is the first and biggest challenge that the embryo faces, and a great many fail to do so and are lost with the monthly flow.

Ten days should be plenty of time for the embryo to make it to the uterus, but the problem is that embryos, like eggs, move along aimlessly. Occasionally, an embryo will not make it out of the fallopian tube and into the uterus by the time it reaches the 256-cell stage. When this happens, the embryo tunnels into the tube walls just as it would into the uterus. This is an ectopic pregnancy. Through the first eight weeks of pregnancy, embryos are incredibly tiny, and their need for nutrients and oxygen is served perfectly well by simple diffusion from the surrounding tissue. Thus, neither the embryo nor the fallopian tube will initially notice anything amiss during the early stages of an ectopic pregnancy. However, as the embryo continues to grow, the trouble starts.

The fallopian tube is in no way equipped to support a pregnancy, and the embryo becomes like a parasitic invasion. The embryo itself has no way to detect the problem, and it continues on its aggressive program of expansion and development. The fallopian tube, unlike the uterus, cannot cleanly abort the doomed and increasingly dangerous pregnancy. Eventually, the untenable situation comes to a head as the growing embryo presses against the walls of the fallopian tube; this will likely be the first time the woman senses that something is not right. The pressure will become increasingly painful, and if she does not seek medical intervention, the embryo can rupture the fallopian tube. Severe pain and internal bleeding follow, and without emergency surgery to repair the damaged tissue and seal the bleeding vessels, the woman may bleed to death, killed by her own offspring attaching where it should not.

There is an even rarer, stranger, and more dangerous form of ectopic pregnancy. Very seldom, when an egg is ejected from the ovary, it does not make it into the fallopian tube at all. This is because, quite oddly, the fallopian tube is not actually connected to the ovary. Rather, the opening of the fallopian tube envelops the ovary, sort of like a too-large garden hose resting on a too-small spigot. The two are not actually attached, and sometimes an egg gets squirted out of the ovary and into the void of the abdominal cavity instead of into the fallopian tube.

When this happens, it is usually of no consequence. The egg simply dies after a few days and is resorbed by the peritoneum, the thin wall of highly vascular tissue surrounding the abdominal cavity. No problem.

However, if an egg gets squirted into the abdominal cavity and sperm bursts onto the scene within a day or so, it might find this egg and fer-

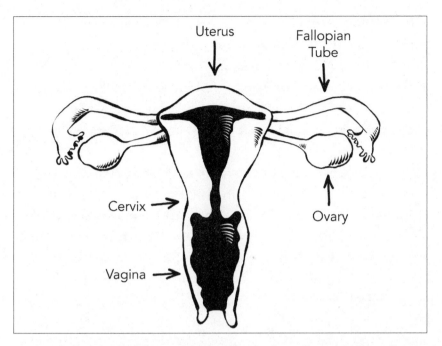

The human female reproductive organs. Because the ovary is not physically connected to the fallopian tube, there is no guarantee that a released egg will even end up in the reproductive system.

tilize it. Once again, this is a rare event because the sperm would have to be searching throughout the lower abdomen for the egg rather than staying in the confined space of the oviduct as it generally does. But this does occasionally happen. The resulting embryo, completely unaware of how far it is from home, mindlessly begins the process of growth, division, and tunneling into whatever nearby tissue that it can find, usually the peritoneum but occasionally the outer covering of the large or small intestine, liver, or spleen.

Abdominal pregnancies pose serious risks. In developing countries, they usually result in the death of the mother. In developed countries, they are easily spotted with ultrasounds and treated with surgical intervention to remove the doomed embryo and repair any damaged tissue or bleeding.

Incredibly, in a small handful of cases, an abdominal embryo makes it into the twentieth week of gestation without killing the mother, is delivered extremely premature by surgery, and survives, though not without serious medical and developmental complications. Though they are invariably termed "miracle babies" by the popular press, these infants survive due to intense and superb medical intervention and a great deal of luck.

Perhaps the opposite of an ectopic "miracle baby" is something called a lithopedion. Occasionally, an abdominally developing embryo will make it into the second trimester before dying and, somehow, not kill or harm the mother in the process. At this point, the fetus is too large to be resorbed by the peritoneum but it obviously cannot be delivered in the normal way as miscarriages or stillbirths are. It's stuck there. The mother's body then reacts to the fetus the way it does to any foreign body that might cause infection: it calcifies the outer layers of the amnion and fetus, covering it in a hardened shell.

The lithopedion, commonly referred to as a stone baby, is very rare; only about three hundred cases have been documented throughout history. These lithopedions usually cause medical problems that must be resolved by surgery, but there have been reports of women carrying them safely and asymptomatically for decades. There is even a report of a woman in Chile who carried a nearly five-pound stone baby in her abdomen for *fifty years*. She gave birth to five children naturally during that time.

While lithopedions and abdominal pregnancies are quite rare, they are also 100 percent the result of poor design. Any reasonable plumber would have attached the fallopian tube to the ovary, thereby preventing tragic and often fatal mishaps like these. Similarly, even the most unimaginative engineer would have given egg cells some sort of means of propulsion or, at the very least, put cilia on the walls of the fallopian tubes to gently brush the fertilized egg cell into the uterus. Either of these would eliminate tubal pregnancies and both are possible with design structures that already exist elsewhere in the body.

But of course nature did not come up with solutions like these, which helps explain why ectopic pregnancies, especially the most common kind (those that take root in the fallopian tubes), are more frequent than you might think; 1 to 2 percent of conceptions result in tubal implantation. This is likely an underestimate, moreover, because at least 10 percent of tubal implantations (and possibly up to one-third) spontaneously resolve through the death of the embryo before it has implanted too deeply, meaning that many surely go unnoticed.

Let's not single out the poor fallopian tubes, though. The entire human reproductive system is littered with inefficiencies and poor design. To recap: Humans mature late, conceal female ovulation, have trouble making healthy sperm and eggs, create embryos that don't implant or that have missing or extra chromosomes, initiate pregnancies that aren't successful—and even when everything goes right, both babies and mothers die in childbirth at shockingly high rates.

In fact, of all the organ systems and physiology in the human body, the reproductive system is the most problematic—the most likely not to work. This is especially odd given how important reproduction is for, you know, *the survival and success of the species.* And it is especially humiliating when you consider that many of these problems are either nonexistent or at least far less common in other animals. When you consider how poorly designed our species' reproductive systems are, it's kind of amazing we made it to the modern era, when science could solve some of these problems for us.

## Coda: The Grandmother Hypothesis

Recently, it was discovered that two species of whales, orcas and pilot whales, undergo menopause. One study found an orca that had died

at the ripe old age of 105, more than forty years after she had last reproduced. So while reproduction in humans is uniquely inefficient and
often uniquely deadly in many respects, when it comes to menopause,
at least, we are not *entirely* alone.

Also called reproductive senescence, menopause is the stage of a
woman's life in which she ceases to have menstrual cycles and is no longer capable of reproducing. And while some whales seem to experience
it, most female mammals go on reproducing through old age, right up
until the end. The cessation of reproduction by a female, even late in
her life, would seem to reduce her chances of passing on her genetic
legacy, so menopause is contradictory to the way natural selection usually works. It is a conundrum that requires an explanation, yet another
possible flaw in our species' reproductive capacity. However, since evolution *did* leave us with menopause, it might somehow be advantageous
to an older woman—or her offspring—that she stop reproducing later
in life. But how?

One idea is that when older women are relieved of the burden of reproducing, they begin to more actively invest in their children and their
children's children, and in so doing, they enhance the success of their
genetic legacy to a greater degree than they would by simply making
more children. But before we explore that possibility, we need to say a
word about how menopause actually happens in women.

I once heard the folk wisdom that menopause is nothing more than
a peculiar byproduct of the recently extended lifespan of humans, the
idea being that, since life expectancy in the premodern era was only
thirty or forty years, women usually didn't live long enough to become
menopausal. Only now that humans are routinely living into the seventh, eighth, and ninth decades has menopause begun to occur.

This idea is predicated on a misunderstanding of what *life expectancy*
really means. While it is true that the average lifespan in the medieval,
classical, and even prehistoric periods was just twenty or thirty years,
the average age of death was low because so many died as infants or as

children. Most people born during prehistory thus didn't make it to reproductive age at all, but many of those who did enjoyed a fairly long life, even by today's standards. We know from ancient texts and recovered skeletons that humans as old as seventy or eighty were not unheard of, even in the prehistoric period. It has been estimated that the average age of death for those who made it past adolescence was somewhere in the late fifties, with lots of people living into their sixties and a few into their seventies.

When it comes to middle age, modern medicine has really added only a decade and a half or so to the end of the human lifespan. The bigger difference has been made in the *first* decade, and that is what has drastically altered the average lifespan average. The point is that women have been living long enough to undergo menopause for hundreds of thousands of years. It is *not* a recent quirk, so we can dispense with that bit of folk wisdom.

Not long ago, it was thought that menopause occurred when women used up all of their ovaries' egg-containing follicles. Women are born with a set number of follicles, around two hundred thousand or so in each ovary. Each of these follicles contains one egg cell, which paused its maturation process way back when the woman was just an embryo herself. Then, each month, anywhere from ten to fifty follicles are selected at random to reactivate the maturation of their eggs in a sort of race. Whichever follicle and egg complete the maturation process first "wins" and is ovulated. The losers all die and, presumably, are never replaced. We used to believe that menopause occurred when a woman simply "ran out" of these egg-containing follicles.

This explanation also turns out to be unsatisfactory, for at least two reasons. First, even if fifty follicles (the high end of the range) are activated every single month and a woman never misses a cycle due to irregular menses or pregnancy, by the time she's in her sixties, she will still have used less than thirty thousand follicles, or one-sixth of what she started with. Second, hormone-based contraceptives interfere with both ovulation *and* the monthly activation of follicles, and thus every year of use should delay a woman's menopause by almost as long. However, women who have used hormonal birth control for decades experience only a modest delay in menopause, if any.

If menopause is not caused by a woman using up all her follicles, what does cause it? It turns out that eventually follicles cease to produce estrogen and progesterone. There are plenty of follicles left; they just sort of peter out. They stop making hormones, and they stop maturing. The symptoms associated with menopause are due to this drop in hormone levels and can thus be treated with hormone-replacement pills. However, this doesn't prevent menopause itself. Sometime in a woman's late forties or early fifties, the ovaries fail to respond to hormonal cues, cease secreting their own hormones, and basically give up.

The precise mechanism of menopause seems to be a timed, age-related decrease in the expression of DNA-repair enzymes in the cells surrounding the egg within each follicle. Without the action of these repair enzymes, DNA damage and mutations accumulate, which accelerates the aging process, and the cells eventually enter a state called senescence. They don't die; they just stop dividing and renewing as follicular cells are supposed to. The ovaries go into a coma of sorts. They are very much alive, but they are permanently idle.

This may sound like a normal, unavoidable, age-related breakdown, like skin losing elasticity or bones becoming brittle — but it's really not. Those and other age-related processes are the result of accumulated protein and DNA damage *despite* the tissues doing everything they can to repair damage as it inevitably occurs; eventually, time wins, and the repair machinery itself becomes damaged and the death spiral of aging sets in. In the case of the ovaries' follicles, however, the genes for the DNA-repair enzymes simply switch off. The aging that happens in the follicle is not slow and cumulative; it is timed and sudden.

This brings us to the evolutionary purpose of menopause. It's easy to conceive that mutations could appear that turn off the ovary's DNA-repair machinery late in life, but why did natural selection favor these mutations instead of eliminating them? The most interesting explanation for menopause is that it allows older women to redirect their efforts toward the success of their children's children. For this reason, this explanation is called the grandmother hypothesis. This hypothesis has been enormously popular—I suspect as much for its explanatory power as for how well it matches up with our cultural notions of doting grandparents spoiling their grandchildren. But the reasoning is actually more complicated than it sounds, because any consideration of the evolutionary value of a certain phenomenon must balance pros and cons.

If an older animal ceases her own reproduction and instead aids her existing children in caring for her grandchildren, it seems obvious that those grandchildren would benefit from that help and be more likely to thrive and succeed. Thus, contributing to grandchildren has obvious natural-selective advantages. However, by forgoing her own reproduction later in life, a menopausal grandmother is reducing the total number of children she can produce. A competitor grandmother who doesn't become menopausal will have more children, and those children will in turn have more children. While they won't have a grandmother's help, they will outnumber those who do. Especially if they show strong kin-cooperative behaviors (as humans do), they might be just fine without the help of a doting grandmother.

So the question is, does a grandmother's contribution to her grandchildren bring so much selective advantage that it is worth the cost of a lower reproductive rate? This thorny question has kept some biologists from endorsing the grandmother hypothesis, and there is one very large and obvious piece of evidence against this explanation: the lack of programmed menopause in other species. If grandparental investment is so great, then those benefits should be seen across many social species, not just humans, yet none of them really do the menopause thing.

One explanation for why the grandmother hypothesis may apply almost exclusively to humans is that the structure of our social groups was, and still is, quite peculiar. All research indicates that, over the past seven million years, our ancestors lived in small, close-knit communities that were highly mobile and socially intricate. There was probably a great deal of experimentation with different lifestyles over that time, as evidenced by the interesting mosaic of anatomical features found in various hominin species. None of that is unique to humans, but one thing may have been: an elaborate division of labor.

As our ancient ancestors became increasingly intelligent and socially sophisticated, they began to amplify the complexity of their already complex primate lifestyles. Tool fashioning, organized hunts, and communal parenting bring a great deal of efficiency to the work of staying alive and free up some individuals to explore and innovate. Before long, early humans were reshaping their world by constructing shelters and fabricating complex tools and manipulating the plants and animals around them. Individuals began teaching skills to one another and dividing labor among the members of the group. It was this environment of communal living that might have provided just the right milieu for the grandmother effect to evolve.

In a highly social group, each member pulls his or her weight in a different way. There is much work to be done. At any one time, some hunt, some gather, some build, some watch out for predators or competitors, some fashion tools, some nurse children, and so on. But just because the individuals live together doesn't mean that they don't also compete with one another. Cooperation helps the group compete better against other groups, but within the group, competition will also be found. In the end, natural selection operates through the success or failure of individuals.

With this in-group competition in mind, imagine a small community with children of all ages. Mortality rates are high throughout childhood, and the children compete for access to food as well as parental care and protection. When a woman is young, her evolutionary interests are probably best served by having as many children as possible to compete against other children for access to resources. Communal parenting means that the burden of childcare is borne by all, so she'd want to try to consume the biggest possible piece of that pie.

As she ages, however, and the number of her offspring rises, her calculus shifts. Her children will end up competing against one another, and her ability to help them all will become compromised by her age and growing infirmity. Success for one of her children may come at the expense of another one of her children rather than at the expense of someone else's child, making it a zero-sum game for her. Continuing to have children may contribute very little to her reproductive potential. In fact, it could actually *hurt* her potential, given how dangerous childbirth is to human mothers. Under these circumstances, shifting her focus toward taking better care of the children she already has rather than making more of them may be a better use of her energy and resources. By this time, of course, her children may have children themselves.

So goes the grandmother hypothesis. It seems a little too neat, but it does fit with common cultural experience and also with some of the unique aspects of human beings: communal living with division of labor, high infant and maternal mortality rates, and long lifespans. This may have been the perfect cocktail of biological factors that rewarded the spontaneous mutations that gave us menopause.

Back to the whales. Researchers analyzed thirty-five years of data, including thousands of hours of video, detailing the movements and activities of orcas living off the coast of British Columbia. What they found was that, as orcas hunt for food in small foraging groups, leading the pack is often an older, menopausal female. In fact, hunting groups frequently consist of an elderly matriarch and her sons. Adult male orcas spend much more of their time hunting and foraging with their mothers than with any other whales, including their fathers.

Even more dramatically, the tendency for hunting groups to be led by menopausal females is most pronounced during periods of famine. When times are tough, orcas turn to their matriarchs, usually their own mothers, to lead them through the darkness. An older orca has been hunting and foraging for many decades and, since whales have impressive memories, she possesses a lifetime of ecological knowledge regarding where to find seals and otters, when the salmon begin their spawning runs, and so on. This knowledge is especially important when food is scarce. It's not clear why older male orcas don't similarly share their knowledge, but the older females definitely do.

Menopause aside, most of the reproductive quirks that humans experience do not seem to be adaptive or shared by other animals. From the late onset of maturity to the eventual arrival of menopause in females, human beings have markedly error-prone and even deadly reproductive systems. These stark reproductive flaws would normally be such a handicap to the success of a species that if the needed fixes didn't evolve, the species would go extinct.

But humans persevered despite these flaws. As we did for our other flaws, we used our big brains to create fixes to circumvent these evolutionary problems. In a way, rather than waiting for nature to do it, we took charge of our own evolutionary destiny. Our creative thinking and collaborative social living helped us scrape by during the earliest years of our species, and then the emergence of language allowed us to accumulate wisdom through the ages and teach the clever tricks to our children. And who among us are the repositories of all that accumulated social knowledge? The menopausal matriarchs we call grandmas.

Humans domesticated plants and animals, invented engineering, and built cities. The advantages that came with these innovations offset our species' low reproductive rates and high child and maternal mortality rates, and eventually this collective knowledge surged exponentially with the dawn of the enlightened scientific era, freeing people (for the most part) from the deadly paradox that reproduction had foisted upon them for so long.

Ultimately, this intelligence allowed us to overcome the limits of biology. Modern medicine has tamed many of the beasts that killed our ancestors so early and so often. Accordingly, as medical standards of care began improving in the mid-nineteenth century, the human population exploded. Along with this explosion came the sinister handmaidens of success: resource scarcity, war, and environmental degradation unlike any our species has ever seen.

So we now have the opposite problem: too many people, rather than too few. Nothing says "poor design" quite like uncontrolled and unsustainable population growth. So maybe all those reproductive limitations weren't so bad after all?

# 5

# Why God Invented Doctors

*Why humans' immune systems so often attack their own bodies; how developmental errors can wreak havoc with our circulation; why cancer is inevitable; and more*

W e humans are a sickly bunch. You may recall from the first chapter of this book that we get head colds far more often than other
mammals, thanks to our peculiar sinus-cavity drainage. But that's just
the tip of the iceberg. Our species is plagued by quite a few other illnesses as well, many of which are particular to us, and many of which
have causes that are much less straightforward than a misplaced drainage hole in the sinuses.

For instance, humans get gastroenteritis a lot, a deeply unpleasant
condition better known (in the United States, at least) as stomach flu.
Gastroenteritis is an umbrella term for any infection or inflammation
in the digestive tract that leads to some combination of nausea, vomiting, diarrhea, lack of energy and appetite, and inability to digest or
even *ingest* food.

These two ailments — head colds and gastroenteritis — are the most
common illnesses in the developed West. While they are rarely deadly,
they nevertheless are so common that they cost *billions* of dollars annually, mainly attributable to wages lost as workers rest and recover.

Tragically, some forms of these diseases are much costlier; for instance, diarrheal diseases (types of gastroenteritis affecting the intestines and, in developing countries, usually caused by sewage-contaminated water) remain one of the biggest global killers.

Neither head colds nor stomach flu nor diarrheal diseases are common plagues among other animals. Of course, although head colds are partially attributable to evolution (owing to our species' poorly designed nasal cavities), they are also infectious processes, just like the stomach bugs that smite us every so often. And when it comes to infectious diseases, humans should blame ourselves first and nature second. That's because diseases such as these are at least partly due to the high population densities and other living conditions peculiar to urbanization.

Beginning in classical times, humans started living on top of one another in booming but filthy metropolises. Their livestock lived on top of one another as well (and still do) and also on top of the humans. Our ancestors' raw and prepared food was in the mix too. These unsanitary conditions—which humankind proceeded to endure for centuries—resulted in a witches' brew of bacteria, viruses, and parasites of all kinds. We can somewhat manage this epic pile-on now, thanks to the invention of modern plumbing. But when you think about the pestilence that our species' way of life invites, it's really kind of amazing that human civilization made it off the ground at all.

All of our ancestors who managed to survive childhood developed antibodies—a type of protein produced by the immune system to protect against bacteria and viruses that might otherwise be fatal. These antibodies made them immune to at least the worst of the bugs that festered in their surroundings. When the age of European exploration began, the aboriginal people with whom Europeans came in contact did not fare well. They hadn't needed the antibodies that European children were forced to develop in order to survive. While they had no doubt developed antibody-based resistance to their own set of infectious agents, they were completely unprepared for the cocktail of pathogens that arrived with the invaders.

For humans today, the infectious ailments that are a routine part of human life were born and bred of the wretched conditions of urban life in Europe and Asia. We therefore can't really call most infectious diseases a design flaw; as I mentioned previously, they are our own fault, not nature's.

Yet we do have design flaws that make us sick. We are plagued by an immune system that seems to be constantly misfiring. When it isn't mistakenly attacking our own cells and tissues in autoimmune diseases, it is overreacting to harmless proteins. Just as humans reach what should be their prime middle-age years, the cardiovascular system starts to develop weaknesses that will only get worse. Not long after, cancer strikes, usually the result of nothing more than accumulated damage occurring within cells.

While none of these conditions are unique to humans, most of them are much more pronounced and deadly in humans than they are in most other animals. We suffer from these illnesses much more than our pets do, more than zoo animals do, and *far* more than animals in the wild do. For reasons that defy logic, it almost seems as if we are built to be sick.

## We Have Met the Enemy, and It Is Us

Of all the diseases that humans have evolved to suffer, autoimmune disorders are among the most frustrating. They do not involve bacteria that we can fight with antibiotics. There are no viruses against which we can develop antibodies. There are no tumors that can be sliced out, poisoned, or irradiated. When we track down the cause of the malady, we find only ourselves.

Autoimmune diseases are the result of mistaken identity. An individual's immune system "forgets" (or never learned) that some protein or cell in the body is its own and not a foreign invader. Not recognizing its own cells, the immune system attacks it vigorously. It's a tragic case of friendly fire.

Predictably, this does not end well. When a body begins attacking itself, there is little physicians can do except give medications that suppress the immune system. This is quite dangerous, and so it must be done very carefully and with close monitoring. There are also all sorts of complications. Besides the obvious threat of infections and a higher rate of common respiratory ailments, drugs that dial down the responsive power of the immune system cause side effects such as acne, trembling, muscle weakness, nausea and vomiting, an increase in hair growth, and weight gain. Long-term use of these immunosuppressants can lead to fat deposits on the face (sometimes called *moon face*), kidney dysfunction, and high blood sugar levels, increasing the chance of diabetes. They also increase the risk of cancer. The treatment can be nearly as bad as the disease.

Nearly all autoimmune disorders strike women more often than men, for reasons no one understands. As if this were not cruel enough, autoimmune diseases often develop slowly and imperceptibly. Patients become accustomed to the pain and limitations and may even doubt that there is anything physically wrong with them. This is compounded when others, even their physicians, brush off their symptoms. A friend of mine has an often debilitating suite of symptoms that come from chronic fatigue syndrome and rheumatoid arthritis, two probably related autoimmune diseases. She has been told by medical professionals, "Well, none of us feel that great first thing in the morning"; "It sounds to me like you need to get out of the house more and get more physical exercise"; and the always helpful "This could just be in your head, but either way, lying around won't help."

Not surprisingly, depression often accompanies autoimmune disease. When your symptoms are debilitating, when you have few treatment options, when you face difficult side effects of treatment, such as acne and weight gain, as well as the looming specter of a life marked by chronic illness, you can become depressed. This is exacerbated when people around you are less than understanding. The lack of support plus the depression often leads patients to withdraw socially, which makes both the physical symptoms and the depression even worse, sending them into a downward spiral of declining health. As my friend puts it, "It feels like I'm drowning, and when I reach out for help, people put weights in my hands and tell me to swim harder."

Autoimmunity is as scientifically baffling as it is heartbreaking. Symptoms can be localized, as in rheumatoid arthritis, which causes painful inflammation in certain joints, or they can be systemic, as in lupus, in which B cells attack other cells everywhere in the body. In both cases, the immune system simply attacks parts of its own body. There's no conceivable reason. This is not some unfortunate evolutionary tradeoff that brings some other benefit. There is no upside to autoimmune diseases. They are simply mistakes. The immune system sometimes misfires.

Autoimmune diseases appear to be on the rise, but, as with other chronic illnesses, it is unclear how much of the increase is due to better diagnostics and extended longevity. The National Institutes of Health estimate that 23.5 million Americans, or more than 7 percent of the population, suffer from one of the twenty-four most common autoimmune diseases. This number is surely an underestimate given that additional autoimmune diseases have already been identified and many more await official scientific classification.

The strangest autoimmune diseases also shine the most penetrating light on this evolutionary flaw. For starters, consider myasthenia gravis (MG), a neuromuscular condition that begins as droopy eyelids and muscle weakness and can progress to complete paralysis — and, if untreated, can ultimately lead to death.

Nothing is actually wrong with the muscles of MG patients. Their immune systems simply begin making antibodies that interrupt normal muscle activity. To get a muscle to flex, a motor neuron releases tiny packets of neurotransmitters onto receptors located in the muscle tissue. The neurotransmitter causes the muscle to contract. It all happens very fast. But if your immune system interferes with the neurotransmitter's receptors, as happens in people with MG, your muscles slowly start to weaken.

The immune systems of patients with MG produce antibodies that attack the neurotransmitter receptors on muscles. Why? No one knows. Fortunately, what follows is not a huge systemic response. If it were, MG would be quickly fatal. What happens is that the antibodies literally get in the way of the neurotransmitters' receptors. As MG progresses, the immune system releases more and more of these antibodies, and patients gradually lose the ability to flex any of their muscles.

Not too long ago, if you had MG, you would die within ten years due to the eventual inability to expand your chest and breathe. Fortunately, myasthenia gravis has become one of the many success stories of modern medicine. In the early part of the twentieth century, the mortality rate of MG was around 70 percent. Today, it is well below 5 percent in the developed West. A series of treatments have been developed over the past six decades, culminating in the current regimen of immunosuppressants combined with special drugs that counteract the effects of the bad antibodies.

This treatment is no picnic. In addition to causing side effects, the inhibitors must be taken at precise intervals, often meaning patients must wake in the middle of the night to take a pill. Many of them must do it every night for the rest of their lives. If they sleep through their alarms because they are sick, had a little too much to drink, or are just exhausted, they will likely have flare-ups of MG symptoms the next day. Even the most careful patients must contend with the occasional crisis, which often requires hospitalization.

Around sixty thousand people suffer from MG in the United States, and it is slightly more common in Europe for some reason. Like most other autoimmune diseases, there aren't even any real hints as to the cause. The immune system simply screws up, and once it starts making the antibodies, it can't stop. Although a genetic form of the disease has been found, it is very rare. The vast majority of cases have no explanation other than some sort of design flaw in our species' immune system. Thankfully, science now saves the lives of most MG sufferers, but for thousands of generations before the previous one, it was a terminal illness.

Like MG, Graves' disease is an autoimmune disease caused by the immune system developing antibodies to a perfectly normal, abundant, and important molecule in the body. In Graves' disease, for no apparent reason, the patient begins making antibodies that act on the receptor for a hormone called thyroid-stimulating hormone (TSH). As its name implies, TSH is the master hormone of the thyroid gland; it induces the gland to release thyroid hormones (THs). These hormones travel throughout the body and have myriad effects, mostly related to energy metabolism. Nearly every tissue has TH receptors, which explains why these hormones have such varied effects in such disparate parts of the body.

In Graves' disease, the antibodies to the TSH receptor do something quite strange. Instead of blocking the receptor and turning it off, they actually *stimulate* the receptor, probably by mimicking TSH itself. In so doing, they prompt the thyroid to release its eponymous hormones.

Normally, the body closely monitors the amount of THs being released by the thyroid. But in the bodies of people with Graves' disease, the thyroid is pounded with the TSH-mimicking antibodies. It responds by releasing ever-increasing amounts of THs, leading to a condition called hyperthyroidism.

Graves' disease is the most common cause of hyperthyroidism. Its symptoms include rapid heartbeat, high blood pressure, muscle weakness, trembling, heart palpitations, diarrhea, vomiting, and weight loss.

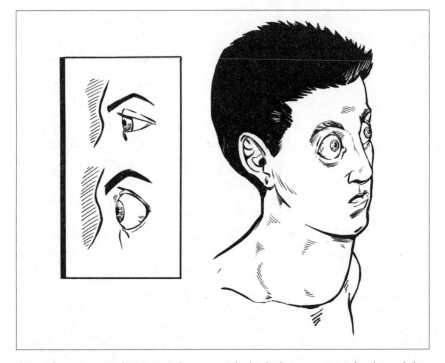

Face of a person with Graves' disease, with the bulging eyes and enlarged thyroid gland (goiter) characteristic of this mysterious autoimmune disorder. Before modern science identified treatments for this condition, many of its victims ended up in sanitariums, suspected of demonic possession.

Most patients develop visible goiters, and their eyes become excessively watery and may even bulge; babies born to women with hyperthyroidism suffer a higher rate of birth defects. The patient may also have psychiatric symptoms such as insomnia, anxiety, mania, and paranoia, and severe cases can result in psychotic episodes. A relatively common condition, hyperthyroidism usually begins after age forty and affects around 0.5 percent of men and 3 percent of women in the United States.

Prior to this disorder being described in 1835, it seems likely that undiagnosed Graves' disease was frequently fatal. It is easy to imagine that the psychiatric symptoms, along with bulging eyes and goiter, might have led our extremely superstitious forebears to suspect demonic possession. Indeed, many histories of sanitariums in Europe in the Middle Ages include tales of paranoid patients with growths on their necks and bulging eyes. Many of these were likely Graves' patients, previously healthy and productive, abandoned by their family and peers to live their last years in agony.

Mercifully, modern medicine has brought a host of effective treatments for Graves' disease that usually do not require immunosuppressants. There are several drugs that can be used to inhibit the thyroid gland. There are also drugs to counteract the most threatening symptoms, such as beta-blockers to slow the heart and reduce blood pressure. These treatments do not have many difficult side effects. In addition, radioactive iodine can destroy part of the thyroid gland; this treatment can be repeated if needed. Finally, surgery to partially or completely resect the thyroid can reverse the condition. This must then be followed by thyroid hormone supplementation, easily absorbed with a once-daily pill. Thus, Graves' disease can now be viewed as an example of science triumphantly solving a problem that our own bodies cause—although for countless generations of humans, the story wasn't so rosy.

If modern medicine has all but triumphed over some autoimmune diseases, like Graves' and MG, another condition in this category—lupus—remains incurable and almost completely shrouded in mystery. Officially called systemic lupus erythematosus, lupus can affect a nearly endless list of tissues in the body, causing a whole assortment of symptoms that can vary widely among patients, from muscle and joint pain to rashes and chronic fatigue. In fact, many scientists consider lupus a collection of related diseases rather than a single disorder. Estimates vary, but in the United States, at least three hundred thousand and as many as one million people suffer from lupus. Autoimmune diseases discriminate by sex, and lupus is no exception; women are four times more likely than men to suffer from lupus.

While the actual cause of lupus is poorly understood, the initial trigger is believed to be a viral infection. What kind of virus it is—and why the infection screws up the immune system permanently—is anyone's guess. What we do know is that B cells, the antibody factories of our immune system, begin to create antibodies that target and attack proteins inside the nucleus of its own body's cells. The immune system, in short, begins to wage war against itself.

When the B cells begin attacking themselves, they undergo a reaction called apoptosis, or programmed cell death. Apoptosis is a controlled form of cellular suicide in which cells dismantle themselves slowly and carefully so as to not trigger panic in surrounding cells and neatly package all of their recyclable materials for their neighbors to absorb. Apoptosis is crucial for embryonic development, cancer defense, and the general health and maintenance of tissues, but it is also a key way that the body's cells protect other cells from viruses. When a cell senses that it is infected, it kills itself by apoptosis in the hope that it will take the virus down with it, thus sparing the rest of the organism. In most contexts, apoptosis is a rather beautiful example of the poetry of life: cells selflessly sacrificing themselves for the good of the being.

It's not so poetic in lupus. When the B cells start killing themselves in large numbers, they overwhelm the body's ability to clear the debris effectively and safely, and they begin to pile up. Compounding this mounting problem, B cells in this activated state are "sticky" due to some receptors on their surface that are designed to seek out infected cells and cling to them. The dying B cells tend to form clumps of cells and cell fragments. This recruits other kinds of immune cells that try to engulf and clear the debris. These immune cells, while attempting to help, sometimes get pulled into the mess. The result is a chain reaction of inflammatory responses that occur throughout the body, focused mostly in the lymph nodes and other lymphatic tissues such as the spleen.

That's the clinical version of what happens. The simplified version goes more like this: Lupus patients feel like crap pretty much all the time.

Because these microscopic clumps can get caught just about anywhere in the body, lupus patients suffer a long list of symptoms that can change over time. The clinical symptoms of lupus are pain, which can be in specific muscles or joints but also more broadly in the torso or head; fatigue, which can be episodic or chronic; swelling, which can be restricted to extremities or present as generalized water retention; fever; skin rashes; oral ulcers; and depression. Most symptoms are caused by the sticky clumps of cellular debris that get lodged in unfortunate places, such as the microscopic filter system of the kidney, the gas-exchange sacs of the lungs, and even the pericardium, the fibrous sac around the heart. These clumps do more than just gum up the gears of the particular tissue. When they get stuck, they are still engaged in an active inflammatory reaction, which can then spread to the nearby tissue. Once again, it's just an absolute mess of autoimmunity.

Lupus is especially frustrating in its prediagnosis stages because the patients' symptoms change, which makes it difficult for their doctors to identify the illness, and the patients often lose confidence in their ability to accurately identify and report their problems. Lupus patients are frequently labeled with a whole host of misdiagnoses, including and especially psychiatric ones. *You were complaining of chest pain, but now it's joint pain? Now it's something else again? Maybe what you need is a psychiatrist.*

Well, maybe yes. As with other autoimmune diseases, lupus is often accompanied by a range of psychiatric symptoms, including anxiety, insomnia, and mood disorders. These stem mostly from the headaches, fatigue, chronic pain, confusion, cognitive impairment, and even psychosis that can accompany lupus. One study found that 60 percent of women with lupus were also clinically depressed. Given all the challenges they face, I'm surprised it's not 100 percent.

Just as the symptoms of lupus vary widely, so do the treatments for it. While nearly all lupus patients take one immunosuppressant or another, this must be coupled with medicines specific to the unique manifestation of lupus that each patient has. Lupus patients can expect years-long experimentation with different medical regimens in search of the combination that works best—which can then suddenly stop working for no reason.

Fortunately, the prognosis for patients with lupus has improved steadily over time—and when it comes to lupus, the time frame is long indeed. The disease has been called lupus since the twelfth century, but descriptions of the disorder date back to the classical period. It has been recognized as an autoimmune disease since the 1850s, but definitive laboratory tests eluded scientists for a hundred years. Today, lupus patients have a life expectancy that is nearly the same as the general public's. But this comes at great cost. There is no such thing as a symptom-free day with lupus, and flare-ups can render patients bedridden for weeks at a time.

It's hard to see lupus as anything other than poor design. Our species' immune system has checks and balances to ensure that the body mounts a vigorous response against foreign cells and proteins while leaving its own cells and proteins alone. During a viral infection, some of the restrictions are temporarily loosened so that the body can more aggressively fight the virus that has hijacked its cells. With lupus, the switch never gets reset, and patients live the rest of their lives in a fight against a phantom virus. The response itself is preprogrammed and useful in its proper context. It is the switch that fails. While all autoimmune diseases are rough, lupus is arguably the most baffling. When the immune system fights itself, it loses either way.

Myasthenia gravis, Graves' disease, and lupus are just three of the many autoimmune diseases that humans can develop. While the National Institutes of Health tracks only the twenty-four most common ones, such as rheumatoid arthritis, inflammatory bowel disease, myasthenia gravis, lupus, and Graves' disease, the American Autoimmune-Related Diseases Association estimates that more than one hundred autoimmune conditions exist, and they affect fifty million

Americans, or about one-sixth of the population. Some of the other diseases that are confirmed or strongly believed to be autoimmune in nature are multiple sclerosis, psoriasis, vitiligo, and celiac disease. Many also suspect that autoimmunity underpins at least some cases of type 1 diabetes, Addison's disease, endometriosis, Crohn's disease, sarcoidosis, and many others. There are hundreds of ways that our immune systems can go wrong and end up making us very, very sick.

To be fair, humans do share a few of these autoimmune diseases with other species. For example, dogs are known to get both Addison's disease and myasthenia gravis. Both dogs and cats can also get diabetes. Interestingly, these diseases are much more common in domesticated animals than in wild animals. We have no clue why the wild cousins of domesticated species and our own close relatives the apes are not heavily burdened with the scourge of autoimmune diseases.

To date, no syndrome similar to lupus has been described in any species other than humans, not even domesticated animals. The same goes for Crohn's disease and many other disorders. While biomedical research has managed to create animal models for some autoimmune diseases, they don't appear to be common in other animals. When it comes to autoimmune diseases, humans and our companions seem to be sicker than wild animals, and we don't know why.

Don't get me wrong: The human immune system is a marvel. It has an overlapping array of defensive cells, molecules, and strategies that keep most of us healthy day to day. Without the immune system, we would succumb to invading bacteria and viruses in no time. To call the immune system poorly designed would be an insult to the millions — nay, *billions* of battles that it wins for us each and every day of our lives.

But to call our immune system *perfectly* designed would be equally inaccurate. There are millions of people who once happily walked this planet only to meet their demise because their bodies simply self-sabotaged. When bodies fight themselves, there can be no winner.

## Overreact Much?

It seems that almost everyone is allergic to something these days. And as anyone with a severe peanut sensitivity can tell you, not all of these allergies are created equal. There are rather innocuous allergies that cause mild flulike symptoms—or an itchy tongue in the case of some food allergies—but allergies can also be lethal. In 2015 in the United States, at least two hundred people died from food allergies; over half of those deaths were caused by peanuts. Tens of thousands more were hospitalized.

While allergies are not quite as baffling as autoimmune diseases, the two conditions share a common thread: In both, the human body's immune system simply gets it wrong. But unlike an autoimmune disease, in which the body overreacts to itself, an allergy is the result of the immune system overreacting to a foreign substance—one that is totally harmless.

Any molecule that triggers an immune response is called an *antigen,* and antigens are usually proteins. Antigens can be found anywhere and everywhere. Everything we eat, touch, and inhale contains potential antigens, but nearly all foreign substances that we encounter are completely benign.

If we couldn't differentiate the innocuous proteins from the dangerous kind, we'd be allergic to everything, but fortunately, the body can usually tell the harmful and nonharmful molecules apart. When a foreign protein is harmless, the immune system generally ignores it. When, however, it's a harmful bacteria or virus, the immune system mounts an attack in order to neutralize the invaders. Such an attack is known as an *immune system response*—a misleadingly innocuous phrase.

One of the principal phenomena of an immune response—and one of the key mechanisms in allergies—is something called inflammation. There are two types of inflammation, systemic (whole body) and localized, and they share some features. The four main characteristics of inflammation have been known since classical times and are still often taught with their Latin names: *rubor* (redness), *calor* (heat), *tumor* (swelling, also called edema), and *dolor* (pain). You will easily recognize these four features in an infected cut, but they are also present during a systemic immune response, such as when you have the flu. You will be flushed (*rubor*) and feverish (*calor*); you will possibly have fluid (*tumor*) in your lungs; and your whole body may ache (*dolor*).

Many of these same symptoms appear during an allergic response, which demonstrates that these symptoms are not due to the infectious invader per se. Rather, they are the work of the immune system in fighting the invader. The redness and swelling are the result of blood vessels dilating and becoming more leaky, which accelerates the process of delivering immune cells and antibodies to the sites of infection. A fever is mounted in an effort to inhibit bacterial growth. Pain is the body's way of nudging you to nurse and protect an infected wound or, in the case of a systemic infection, to lie down and rest, conserving energy for the immune fight. All of the symptoms of inflammation are the result of your body's attempt to fight whatever is ailing you.

Inflammation is definitely beneficial when fighting an infection but it is totally unhelpful in the case of allergies. An allergic antigen, say, the oil from a poison ivy plant, poses no actual threat to the body. Mounting an immune response to poison ivy oil is downright silly. Yet most of us do exactly that whenever we come in contact with it.

Stop and think for a second about how ridiculous allergies are. Some people's bodies go so crazy over a bee sting that they die. The bee stings don't kill them; their immune systems do. Even if bee stings were truly dangerous (which they're not), suicide still seems like an overreaction. Because of hypersensitive allergies, some people's immune systems are like ticking time bombs. The biggest health dangers they'll ever face in life is right inside them.

One of the main culprits in allergic responses is a specific type of antibody that is normally used only to fight parasites and thus is one of the least commonly used antibody types, at least in the developed world. This antibody's main function is to induce and maximize inflammation. For some reason, this parasite-fighting antibody gets released during an allergic response and that's why the inflammation that occurs during an allergic response is so much worse than a standard inflammatory response. Inflammation is all that this antibody knows how to do. When you're a hammer, everything looks like a nail.

Allergies are a conundrum because we are all bombarded by foreign material all the time. We eat food from a variety of plant and animal species. We inhale pollen, microbes, and particulates from a whole variety of sources. Our skin comes in contact with a whole host of substances, including clothing, soil, bacteria and viruses, and other people's bodies. We contend with that onslaught of foreign material just fine, but if a person has a peanut allergy and so much as tastes some peanut butter, he might find himself fighting for his life.

So how does the body tell the difference sometimes and not others? We still really don't know. One thing we do know, however, is that the body needs practice to do this correctly and the environment in which it practices does matter. The training of the immune system takes place in two phases, first in utero, then in infancy.

A fledgling embryo develops immune cells while in utero. The very first thing that these cells do is participate in a phenomenon called *clonal deletion*. Clonal deletion is the process by which the developing immune cells in a fetus are presented with small bits of chewed-up proteins from the fetus's own body. The immune cells that react to those bits of self-protein are then eliminated; they are "deleted" from the immune system. This process goes on for weeks and weeks, and the goal is to eliminate every single immune cell that has the potential to react to its own body. Only then is the immune system ready for action.

It's okay that the immune system is not functional prior to birth because, while the womb is not perfectly sterile, it is pretty close to it.

In this safe environment, the fetus attempts to entrap its own immune system: it dangles little bits of self-antigens, and any immune cells that pounce on it are killed. The result is an immune system whose cells attack only foreign cells. Not long before birth, those cells are activated and the fetus is ready to face a dirty world full of microscopic danger.

Once infants are born, the challenge gets more difficult. As the baby bursts into the very septic world, his immune system is bombarded by antigens it has never seen before. It has to learn who is friend and who is foe, and quickly. From a newborn's first day of life, its immune system faces various infectious agents that it doesn't yet know how to deal with, some mild, some serious. How does the body know to fight one strain of *Staphylococcus aureus* with everything it has but ignore another strain? No one really knows. One thing seems certain: the early immune system reacts slowly and adopts a "wait and see" approach.

Many scientists think this is the key to phase two of immune training—the body figures out which foreign proteins are dangerous and which are harmless by going slow with the immune reaction at first and seeing if an infection takes hold. If it does, it's time to kick things into high gear; if it doesn't, the foreigner is seen as a big deal. The immune system has an incredible memory, as evidenced by the fact that vaccines to now extremely rare infections are still effective decades after people received the vaccinations. But initially it must learn who is friend and who is foe, and there is simply no other way to learn except firsthand experience.

The result of this slow immune response is that truly dangerous infections get a head start on infants. Any parent will tell you that kids are constantly sick. Part of this is because they're still building immunity to viruses, such as those that cause chest and head colds, but another part is that their immune systems are learning what bugs to fight and how to fight them. When the immune system does decide to jump into action, it usually does so very strongly, which compensates for the late start. This is why children tend to run much higher fevers than adults do. I once measured a fever of over 106°F in my son for nothing more serious than strep throat (though at the time, racked by the jitters of a first-time parent, I assumed he had the plague). If my temperature goes above 101°F, I feel like I'm dying.

Importantly, our immune systems learn to be tolerant of the daily grind of life on earth. Most of the foreign molecules in the air, in our food, and on our skin are completely harmless. The majority of bacteria and viruses are harmless too. Our immune systems get used to the constant barrage of foreign material and learn not to fight it. Beginning at just a few months of age but continuing for the first few years, the immune system starts to settle into a mature state, assuming that it has seen most of the harmless stuff by then.

As it transitions out of the infant learning phase, however, the immune system begins to change. It becomes more sensitive to new foreign material that it comes in contact with. This is when allergies rear their ugly head. Instead of learning that a harmless substance like peanut oil poses no health threat, the immune system decides to fight it, a reaction that will become more potent with increasing exposure. In other words, the immune system learns the exact opposite lesson than it should.

There is no evolutionary explanation for why we get allergies, and all animals can suffer from them. However, as with autoimmune disease, no species suffers from allergies as much as humans do. Prevalence of both food and respiratory allergies has been skyrocketing in the past two decades, and currently over 10 percent of children in the United States have at least one food allergy. When I was in elementary school in the early 1980s, I didn't know a single kid with a peanut allergy except my sister, who was eleven grades ahead of me. Nowadays, both of my children usually have multiple kids in their classes every year who are deathly allergic to peanuts or other nuts. Many schools and daycares have opted to go entirely nut-free rather than deal with the constant worry of protecting the allergic kids from the scourge of nuts that could send them into anaphylaxis. Knowing what we know about how the immune system is trained and what goes wrong in the development of an allergy, what has changed over the past few decades to send allergy rates through the roof?

The likely answer is something called the hygiene hypothesis. Beginning in the 1970s and '80s, people started going to great lengths to minimize children's, especially infants', contact with germs. Today, parents sterilize their babies' bottles and ask visitors to wash their hands before holding or touching them. They keep infants mostly indoors and definitely off the bare ground. Only the cleanest food and liquid for their tummies and always freshly washed clothes for their bodies. If the pacifier falls on the floor — *Stop! We must sterilize that now!*

This is all very well intentioned and it's hard to argue with any of these daily decisions. I have instructed my children in no uncertain terms never to eat anything off the floor, to avoid using public bathrooms, and to touch nothing while riding the subway. I insist on these precautions because I don't want them to get sick.

Furthermore, if you have a cold, it just seems like common sense that you shouldn't hold a two-week-old infant. In some circles, it is even considered a faux pas to visit someone with a newborn if you have young children. It doesn't matter if you leave them at home; you might have germs on your clothes and your person that could make the infant sick. Again, this is a well-intentioned protective parental reaction.

Good intentions aside, when safeguards like these are taken to extremes, they unwittingly wreak havoc with how evolution has shaped the development of our immunity.

It turns out that the sterilization of infant life may be the reason behind the rise of allergies. Several studies have now implicated an excessively clean environment during infancy in the development of food allergies later. This is the hygiene hypothesis. It makes a whole lot of sense because the one thing we know about immune system function is that it requires a lot of practice to work well. This is why most vaccines are not given to children immediately at birth. Their immune systems just aren't ready. It's not that vaccines harm infants; they just don't work. The same principle applies in the other direction — minimizing exposure to antigens will prevent children's immune systems from getting accustomed to them. Only by seeing a lot of both harmful and harmless foreign substances can our immune systems learn to tell the difference.

If this hypothesis is correct, we are collectively taking a relatively minor design flaw — allergies — and blowing it up to epic proportions. For that, we couldn't blame nature. The fault would be ours.

## Matters of the Heart

Cardiovascular diseases are the number-one cause of natural death in the United States and Europe. Collectively, in fact, coronary artery disease, stroke, and hypertension are the ultimate cause of about 30 percent of deaths in the developed West. Most of these fatalities are attributable to problems with the heart itself, but dysfunction of blood vessels are also frequently to blame. (Most kidney diseases, for instance, are actually circulatory problems that happen to occur in the kidneys because there are so many blood vessels concentrated there.)

Some heart disease is age-related or the result of poor lifestyle choices; if you live long enough or behave unhealthily enough, you are likely to suffer from these cardiovascular conditions. This isn't exactly a design flaw. We really have nobody to blame but ourselves — and odds are, you have already heard plenty on this topic and don't need to hear more about it now. (Surprise: you should eat healthy foods and get plenty of exercise!)

But humans do face some unusual design defects when it comes to matters of the heart. For example, every year in the United States alone, about twenty-five thousand babies are born with a literal *hole* in the heart.

The clinical term for a hole in the heart is a *septal defect,* and it can occur between the two upper chambers of the heart or the two lower

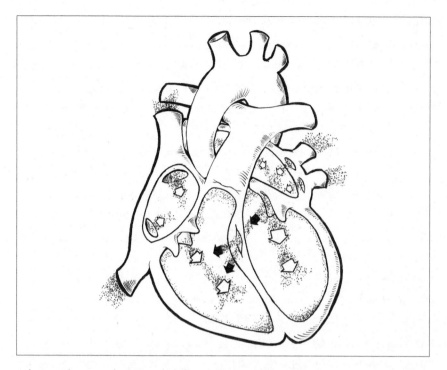

A human heart with a septal defect: a hole in the septum that allows blood to flow from the left side of the heart to the right side. This common yet life-threatening birth defect suggests that the genes governing the development of the human heart are less than fine-tuned.

chambers. When this happens, blood sloshes between two chambers that are normally not connected to each other in the sequence of blood flow. During the heart's contraction, the hole allows blood to flow from the left side of the heart to the right side. When the heart is resting, blood may inadvertently flow from the right side back to the left. The hole results in an improper mixing of venous and arterial blood.

Normally, when blood returns from delivering oxygen to the tissues of the entire body, it enters the right side of the heart. From the right side, the blood is propelled to the lungs, where it picks up oxygen and unloads carbon dioxide. Then the blood returns to the heart, this time to the left side, where it is repressurized and pumped out to the body.

This two-step process is important because blood has to be pumped out at high pressure in order to flow out to the body, but it has to circulate at low pressure so the tissues have time for the gas exchange that is the whole purpose of blood to begin with. Pump out, gas exchange (lungs), pump out, gas exchange (whole body). That is the pattern.

When there is a septal defect, however, blood mixes between the two steps. This is like a short circuit of the normal flow of blood. A tiny hole makes no difference at first, though it may grow over time due to the friction of blood flowing through it. A large hole can disrupt blood flow so completely that it is lethal, either in utero or shortly thereafter. The bottom line is that inefficiencies caused by septal defects place an additional load on the heart. It must work that much harder to circulate blood properly.

Currently, clinical outcomes for children born with septal defects are pretty good. Many defects are so small that no intervention is necessary (though regular checkups are called for). Larger ones must be repaired surgically, a procedure that became an option only in the late 1940s. The septal walls are deep inside the heart chambers, so this means open-heart surgery. This is about as invasive as it gets and requires a complete heart-lung bypass during the operation. It carries all kinds of risks. Nevertheless, doctors have refined the surgery to such a degree that in developed countries, almost all children born with septal defects now survive and live completely normal lives.

This obviously would not have been so just a few decades ago. Severe septal defects were once a substantial cause of immediate postnatal death. If a baby had a gaping hole in her heart, she usually lived for just a few hours, gasping for breath, before slowly suffocating due to the inability to properly circulate oxygen.

Of course, most of us do not have holes in our hearts, and the frequency with which this developmental error occurs indicates that the genes responsible for the genesis of the heart are a little rusty. While septal-development defects are sporadic, they are not due to sporadic *mutations* but rather sporadic failures in the embryonic development of the heart. It's just sort of bad luck, but there seems to be a predisposition for this very specific type of bad luck.

To understand how someone can be predisposed to experience a particular problem, consider your shoelaces. If your shoelaces are tied properly, the odds of you taking one hundred steps without tripping are pretty good, but not zero. If your shoelaces are untied but pretty short, you might still take one hundred steps without tripping, and if you do trip, you probably won't do so more than a few times. If the untied laces are very long, however, you will almost certainly trip multiple times within the one hundred steps, but still, you most likely won't trip every single step.

As this example demonstrates, the odds of a problem — tripping — can be low or high depending on a variety of factors. There is no perfect situation in which tripping is entirely impossible, nor is there a situation where tripping every step is guaranteed. There is just a range of probabilities.

The influence of genes on development is akin to the influence of shoelaces on tripping. There is a low chance of a baby being born with a hole in his or her heart. However, the fact that in the United States alone, a couple of thousand babies a year are born with holes in their hearts indicates that the genetic shoelaces are untied. Somewhere in the genes for heart development, some things are not quite what they should be. The laces may be short, but they are definitely untied.

If you think that's weird, consider this: Some babies are born with the blood flowing in the wrong direction through their circulatory systems. This is a severe problem that must be corrected immediately. Circulation is a closed system, so, in principle, flipping things around in the circuit would still result in blood going to the right places: being refreshed with oxygen in the lungs, sent to the tissues, returned to the lungs for more oxygen, and so on. However, it cannot run effectively in reverse because both the vessels and heart muscle are configured to meet the needs and pressures of different systems. The right side of the heart is built to pump blood only to the lungs and back out to the heart, and it is not strong enough to push blood through the whole body. In addition, the pulmonary arteries, which normally carry blood to the lungs, are built very differently than the aorta, which normally carries blood to the whole body. If their roles are reversed, neither will perform its function very well.

In a dramatic triumph of medical science, some children with this condition—known as transposition of the great vessels—can now be saved. Surgeons must slice out several of the vessels and swap them around in order to match their strength, thickness, and elasticity to the load they must bear when the blood is flowing correctly. This has to be done while the infant is on total heart-lung bypass, so it is incredibly risky to perform on a baby that is just hours or days old. Nowadays, most children do survive the procedure and live relatively normal lives. What nature has goofed up, science can now correct.

While holes in the heart and transposed vessels are life-threatening but rare defects in the formation of the cardiovascular system, there are subtler malformations that are much more common—and that can be every bit as dangerous. One example is anastomoses, bizarre configurations of blood vessels in which fairly large arteries form short circuits with veins to create a futile cycle of circulating blood. These worthless blood vessels can actually pose a lethal threat if they grow large enough. Because they pointlessly receive a great deal of blood flow, even a minor injury to the engorged vessel can result in a massive loss of blood very quickly.

Though many are innocuous, anastomoses do not resolve on their own. An anastomosis that is actively growing must be removed before the mass poses serious health risks. Some of the most dangerous anastomoses form branches and eventually become tangled webs of inter-

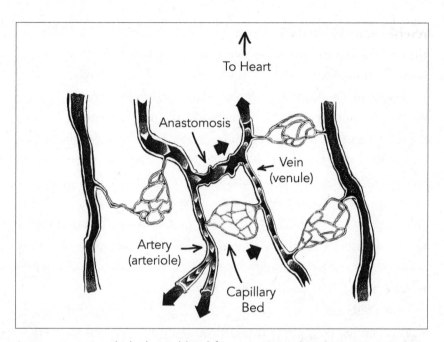

An anastomosis, which shunts blood from an artery directly to a vein without passing through a capillary bed. This leaves the surrounding tissue devoid of oxygen, which can cause the anastomosis to grow in a runaway cycle.

woven vessels. In previous eras, these would have occasionally been fatal and quite often debilitating. Anastomoses, if left alone, tend to grow larger over time, creating an ever-swelling mass filled with stagnant blood. As such, they are usually removed with surgery or destroyed with radiation when they are very small. This correction, however, gets more dangerous the larger the mass grows because the sliced vessels will spew large amounts of blood before clotting can stop it.

Once these insidious structures form, they often end up in a runaway cycle of growth. This is because the tissue surrounding the pointless vessels paradoxically gets starved for oxygen-rich blood. Unlike normal arteries, which convey blood from the heart and branch into gas-exchanging capillaries that transport the precious oxygen to the body's various tissues and organs, anastomosed arteries simply branch directly into another type of blood vessel: veins, which return the blood to the heart. Because anastomoses skip the capillary step, the tissue around them actually becomes oxygen-depleted, a condition called hypoxia, despite the fact that enormous amounts of blood are effectively passing through these tissues every second. In response, the hypoxic cells secrete a hormone that promotes the further growth of vessels in the anastomosis. The vessel gets larger; it may even form branches, and then even more tissue becomes hypoxic, and the cycle continues.

As with so many developmental defects, no one really has any idea how or why anastomoses form; they just do. This is more poor programming in our developmental genes and tissue architecture—basically, the untied-shoelace situation.

## Coda: The Beast That Stalks Us All

Although many people have no allergies, will never have a stroke, and will escape the horror of an autoimmune disease, cancer is the beast that stalks us all. There is basically a 100 percent chance that, if you live long enough, you will get cancer. It *will* catch up with you eventually, provided you don't die of something else.

Rates of cancer in the human population are skyrocketing. This is largely (though not solely) a function of people simply not dying of other things and thus living long enough to get cancer. Furthermore, all species of multicellular animals are subject to cancer. Humans are not unique in this. We get cancer at higher rates than some animals but at lower rates than others.

There is nothing about cancer that makes humans special, in other words, except that we now live so long that we develop it more than we used to. So why mention it at all? Why not skip over it, like atherosclerosis?

Because cancer is the ultimate bug-and-feature of nature. You cannot have sexual reproduction, DNA, and cellular life without also having cancer. Its very ubiquity, in fact, points to its extremity as a flaw of nature—a design defect that affects not only humans but many other living things as well.

Like autoimmune disease, cancer is a product of our cells. It happens when cells get confused regarding their own code of conduct and begin to grow and multiply out of control. The resulting clump of deprogrammed cells, in the case of solid tumors, loses its normal function and strangles the organ in which it appears. In the case of blood cancers —leukemias and lymphomas—the cancer cells crowd out the blood cells and the bone-marrow machinery to make more of themselves. In either cancer type, the cancerous cells usually spread to other tissues and take them over until the body is too crippled to continue. Thus, cancer is essentially a disease of cell-growth control.

Most of the cells of the body are capable of growing, dividing, and multiplying if and when needed. Some cells are almost always growing, like the ones in your skin, intestines, and bone marrow. Some basically never divide, like neurons and muscle cells. And some are somewhere in the middle—not always dividing but capable of doing so during wound healing or tissue maintenance. Cells must thus regulate their own proliferation. They should multiply when needed but stop when it's time to. Cancer begins when a cell ignores the rules and continues to grow incessantly. In this sense, the disease is the corruption of our own cells; it causes them to take on a life of their own, abandoning their proper posts and dedicating themselves solely to their own growth and proliferation.

I was once seated on an airplane next to a Benedictine monk named Father Gregory Mohrman. During our conversation, I mentioned that I was returning home from a conference on cancer research. A very learned man, he was fascinated by this and asked many questions about my research and the nature of cancer itself. He then launched into an eloquent monologue regarding his thoughts on cancer, which I attempt to paraphrase here:

It seems to me that cancer is the ultimate biological manifestation of the devil. Cancer is not the result of a bacterial or viral attack, and it's not that our bodies become damaged by some outside force. It's us. Our own cells, as if seduced by some evil force, forget their proper place in our bodies and begin to live solely for themselves. They become the embodiment of selfishness, taking everything for themselves, sparing nothing for the rest. Never satisfied, they grow more and more and spread to other areas to continue growing and taking and killing. The only way we know to fight these corrupted cells makes us very sick because in attacking the cancer, we attack ourselves. There is no other way to fight the demon that has taken over our very flesh. This is why I have always held oncologists and cancer researchers in the highest regard. You are dedicated to the fight against evil.

The monk's monologue left me breathless, and I have never forgotten it. Ironically, the opening paragraph of any article or text on cancer will often say the very same thing as his poetic yet supremely concise description, although with more clinical—and less interesting—wording. Cancer is indeed the result of nature's poor design—a being's own cells malfunctioning to the point of killing the entire organism. (A notable exception to this is the human papillomavirus, HPV, which can cause cervical cancer. Only a small minority of cancer cases are caused by viruses.)

There are two reasons why cancer is so stubborn. First, as Father Mohrman points out, cancer is not a foreign invader; it is our own cells gone wrong, and so drugs that fight cancer cells while sparing normal cells are hard to come by. Second, cancer is progressive—and usually aggressively so. Cancer cells are constantly mutating, which means that it is not the same disease over time; rather, it grows, morphs, invades, and ultimately spreads all over the body. A treatment that works at first will fail eventually. If a tumor contains ten million cells and doctors kill 99.9 percent of them with radiation and chemotherapy, there are still plenty left to regrow the tumor — and it will be even more aggressive as well as resistant to whatever was used to shrink it originally.

What causes the body's cells to begin growing uncontrollably? It turns out that nearly every cell in the body is subject to occasional mutations, which are random changes to the DNA sequence. Some of these are caused by toxins that we are bombarded with in our environment, but the majority of them are due to mistakes made when cells copy their DNA. With billions of cell divisions taking place daily, they make tens of thousands of errors every day.

This is how most cancer begins. With thousands of permanent mutations occurring every single day, occasionally one of them will hit a gene that nudges a cell away from its proper proliferation control and toward a cancer-like state. Mutations are random. There is nothing very special about so-called cancer genes that make them more susceptible to mutation. Most mutated genes don't drive a cell toward cancer. Some do, however, and when those cancer mutations occur, the cell begins to grow in an uncontrolled fashion.

When this happens, the principles of evolution by natural selection take hold. If a mutated cell grows a little faster than its neighbors, its offspring will outnumber the offspring of its neighbors. The faster growth rate also accelerates mutations, since there is more DNA copying going on and thus more chances for additional errors. Most of those errors will have no effect, but occasionally, randomly, a mutation will occur that drives the cell faster still. That cell will then produce progeny even faster, and its offspring will outnumber the others yet again. Cancer is the result of successive waves of mutation, competition, and natural selection, several of which occur before the tumor is even large enough to be noticed.

Because it is both a bug and a feature of cell division, cancer is largely thought to be an inevitable fact of life for all multicellular organisms. As soon as living things were made of more than just a single cell, the problem of coordinating the proliferation of cells began. Cell division —and the DNA copying that goes with it—is a dangerous game. The more you play, the more likely you'll eventually lose. Unless the human body somehow acquires the ability to flawlessly copy its own DNA —a biological pipe dream if ever there was one—if people live long enough, cancer will strike them at some point in their lives.

The grim irony is that cancer is, in a sense, a necessary byproduct of an essential part of life. Everything great that evolution ever brought was due to mutation. Random copying errors introduce variety and innovation. From an evolutionary perspective, mutations provide genetic diversity, which is good for the long-term survival of a lineage. Mutations in general are thus the ultimate feature/bug system.

Therefore, evolution has struck an uneasy balance with cancer. Mutations cause cancer, which kills individuals, but it also brings diversity and innovation, which is good for the population. Certain species, such as humans and elephants, spend years maturing before they can reproduce, so they must aggressively protect themselves from cancer, lest they succumb before they can have offspring. Shorter-lived species, such as mice and rabbits, can tolerate higher mutation rates and lazier anticancer defenses. True, cancer will eventually get us all, but that is the compromise. Evolution cares little about the individuals who will die of cancer. This is a sacrifice worth making for the diversity that comes from mutations.

As Lewis Thomas put it, "The capacity to blunder slightly is the real marvel of DNA. Without this special attribute, we would still be anaerobic bacteria, and there would be no music."

# 6

# A Species of Suckers

*Why the human brain can comprehend only very small numbers; why we are so easily tricked by optical illusions; why our thoughts, behaviors, and memories are so frequently wrong; why evolution rewards adolescents, especially males, for doing foolish things; and more*

In a book about human weaknesses, it might seem odd to find a chapter about the brain. After all, the human brain is by far the most powerful cognitive machine on the planet. Sure, computers can now beat us at chess and go. But in lots of other regards, we still have a major leg up on machines—even those whose only purpose is to think.

The advancement of the human brain beyond that of our closest relatives over the past seven million years has been truly exponential. Our brains are three times larger than those of chimps, but that doesn't really capture the difference between us because almost all of the growth that the human brain has experienced has been in a few key areas, especially the neocortex, where advanced reasoning takes place. Our advanced processing centers are massively larger and more interconnected than those of any other species. Even modern supercomputers cannot compare to the fast and nimble capabilities of the human brain.

The beauty of the brain is not just in its raw computational power but also in its ability to self-train. Sure, we humans in the developed world subject ourselves to extensive formal education these days, but the most intense and impressive learning takes place outside the classroom. Our species' acquisition of language, a skill far more profound and nuanced than anything we'll ever learn in school, happens naturally and almost effortlessly and is driven purely by the brain's remarkable ability to collect information, synthesize it, and incorporate it into its own programming. Advances in machine learning come nowhere near this level of achievement. Anyone who is reasonably bilingual can easily see just how much smarter the human brain is than a computer by playing around with Google Translate, possibly the most sophisticated translation program publicly available. After just a few months of lessons, the human brain can translate between languages better than the fastest computers can.

But the brain isn't perfect. The human brain is easily confused, tricked, and distracted. There are certain rather low-level skills that it struggles to master. It makes embarrassing blunders even within its otherwise impressive skill set and is beset by bizarre cognitive biases and prejudices that handicap it as it tries—and sometimes fails—to make sense of a complex world. It is overly sensitive to certain inputs and blind to others. And it rigidly adheres to outdated dogmas and superstitions that even the most elementary logic refutes (I'm looking at you, astrology), while a single anecdote can shape its entire worldview on an issue.

While some of the brain's limitations are the result of pure accidents —the unexplained misfiring of a computational instrument with finite capabilities—others are the direct result of how the brain is wired. The power and flexibility of our species' brains evolved while our ancestors were living very differently than modern humans are now. For almost all of the past twenty million years, our species' lineage was that of just another ape. We humans reached our current anatomical dimensions only about two hundred thousand years ago and began the shift toward modern ways of living only about sixty-five thousand years ago. Our species hasn't undergone much genetic change since settling down into civilized life, and so our bodies and brains are built to make sense of a very different world. Our mental abilities—now used for such things as philosophy, engineering, and poetry—evolved for totally different purposes.

The period most crucial to human evolution was the Pleistocene epoch, which began about 2.6 million years ago and lasted until the end of the last ice age, around twelve thousand years ago, a point in time sometimes called the dawn of civilization. By the end of the Pleistocene, humans had spread around the globe, most of the major racial groups were established, agriculture was being developed in many places simultaneously, and the gene pool was little different than it is now.

In other words, human bodies and brains have not changed much in the past twelve thousand years. This is a kind way of saying that we are not adapted to *this* life. We are adapted to Pleistocene life. And perhaps nowhere is this clearer than in the way we perceive the world around us.

## Fill in the Blanks

Optical illusions are a staple of fun houses, museums, circuses, magic shows, coffee-table books, and, of course, the Internet. These visual tricks dazzle us because they leave us with a sense of cognitive dissonance. We know that things aren't quite right as our brains unsuccessfully continue to try to find a solution to the problem. This can be fun but dizzying. Most people get uncomfortable if their brains are confused for too long.

There are dozens of types of optical illusions—physically impossible objects (like a fork that has either three or four prongs depending on which side you look at), perfectly straight lines that look bent or broken, the appearance of depth or movement in a static two-dimensional picture, even spots or images that appear and disappear based on how you move your eyes across them. Each has a slightly different mechanistic explanation, often centering on the theme of our brains' "filling in the blanks" when information is missing (or misleading) in order to create a complete, if inaccurate, picture. The senses relay very raw, unprocessed, nearly unintelligible information, and the brain must construct this mishmash into a coherent picture. It's not unlike the signal that goes to a computer monitor. It's nothing but a rush of electrons that ping out the 1s and 0s of binary code, and yet the video card sorts out the blur and creates a highly organized image.

Unlike a computer monitor, however, the human brain has the fascinating ability to extrapolate from the information it has. This happens unconsciously. Most of the time, it comes in handy. For example, we are highly attuned to faces. Our species has a spectacular diversity of face shapes and structures, and the human brain picks up on these subtle differences instantaneously. While most people struggle with names, the majority never forget a face, and many of us can recognize friends from a single feature, such as an eye or a mouth. This is because faces were key to sociality during the long Pleistocene time before language developed. Humans used faces to recognize one another and communicated with their expressions. This has led to our amusing tendency to see faces in inanimate objects.

As early humans were eking out livings, mental abilities such as drawing inferences from an incomplete picture, predicting future events based on past experiences, and sizing up a situation using only a partial glimpse of it would have been incredibly powerful and often lifesaving. Occasionally, though, this impressive feature of the brain can lead us astray, creating *inaccurate* pictures in our minds.

Entertaining optical illusions exploit these mental faculties. Take, for example, images that appear to move even though they are static. These generally involve alternating or interlocking patterns of some shape with sharp or tapered corners or an otherwise pointy edge. The effect seems to work only when the same patterns are laid out in an opposi-

Patterns of alternating shapes, such as this one, can invoke the sensation of motion in the human brain. This is due to the way our brain creates smooth "video" from the still images captured by our eyes.

tional or alternating fashion. Something about the heightened contrast in the pattern drives the effect. Our brains ascribe movement to these shapes as a side effect of a pretty ingenious neurological innovation shared with many other creatures: the "smoothing out" of perception for objects in motion.

The neurons in retinas capture visual information and relay it to the brain as fast as they can, but this relay is not instantaneous. What we see is not the world as it is now but as it was about one-tenth of a second ago. The delay is due to the maximum frequency that neurons can fire.

This maximum firing rate, when considered for all the neurons in the retina (since they all send information at the same time), leads to something called the flicker fusion threshold: the frequency faster than which our eyes cannot work. When visual information is changing faster than the eyes can detect, the brain "smoothens" this information into the perception of an object in steady motion. In a sense, we do not actually *see* motion; we infer it. The eye takes snapshots—about fifteen per second in dim light—and sends them to the brain. The visual cortex then creates a smooth experience out of what is really an old-time film reel of still pictures.

That is no idle analogy; in fact, much of the visual media that we consume is delivered to us as rapid flashes. Both television and movies have a frame rate, which is the number of screen flashes per second, usually between twenty-five and fifty. As long as this rate is faster than the eyes can work, the brain smooths out the input and creates the perception of fluid motion. If the frame rate were just a little slower, people would perceive television programs and movies as they really are: a strobe of flashing pictures. Part of the reason dogs and cats show little interest in television is that their retinal neurons work so much faster than ours that they actually see the flashes, which must be awfully annoying. Birds tend to have higher flicker fusion thresholds than mammals, helping to explain their impressive abilities to hunt quick prey such as fish and flying insects. Apes and other primates, including humans, have rather slow flicker fusion thresholds despite their superior color vision, indicating that hunting fast-moving prey is not usually a priority. (Humans engage in persistence hunting, which relies on endurance and ingenuity more than quick actions.) Still, human brains do create the illusion of motion out of still pictures, even if we do it more slowly than others.

The same functional anatomy in our brains that creates the perception of smooth motion often misfires when we look at certain patterns. Our brains get tricked like this only with particular shapes. When a person looks at a checkerboard, it doesn't usually evoke the illusion of movement. The patterns that tend to trip our "motion-creating" function are those with sharp corners that seem as though they are pressing forward. In the open plains of the savanna, the motif of a pointy-edged something plunging into an open visual field is reliably associated with movement; our brains are adapted for this.

Artists have long known this and often exploit the brain's ability to create the illusion of motion in their work. A 140-year-old oil painting is as static as can be, but many of Edgar Degas's masterpieces, such as his famous ballet dancers, leave the viewer with the distinct sense that the subjects of the works are in motion.

As error-prone as our visual faculties are, they're by no means our only mental assets to be intrinsically flawed, let alone the most impressively flawed. Our species' advanced computational brain—the single biggest human feature—is full of bugs. These are called cognitive biases, and they can get us into big messes.

## Born to Be Biased

The term *cognitive bias* refers to any systemic breakdown of rational or "normal" decision-making. Collectively, these defects in human decision-making receive a huge amount of attention from psychologists, economists, and other scholars seeking to understand how something as miraculously advanced as the human brain can go so incredibly wrong so incredibly often and with such incredible predictability.

The human brain is, on the whole, a marvel of logic and reason. Even as children, humans are capable of deductive reasoning and learn the simple rules of if/then logic. Mathematics, the basic version of which is an innate skill, is essentially an exercise in logic. This is not to say that reason never escapes us, but in general, humans think and act logically. This is why cognitive biases are strange and call out for study—they are deviations from the rational way that we expect our brains to work.

There is an entire subfield of economics, known as behavioral economics, that has arisen in recent decades to explore these biases. One of the founders of the field, Daniel Kahneman, won a Nobel Prize for this work and has explained many of our biases in his popular book

*Thinking, Fast and Slow.* There are literally hundreds of cognitive biases with overlapping definitions and common root causes, and they're grouped into three broad categories: those affecting beliefs, decisions, and behaviors; those affecting social interactions and prejudices; and those associated with distorted memories. Cognitive biases in general are the result of shortcuts that the brain takes in making sense of the world. In order to avoid having to thoroughly analyze each and every situation you find yourself in, your brain establishes rules based on past experience that help you make quicker judgments. Saving time has always been a priority, and the brain has evolved to save time whenever it can. Psychologists refer to these time-saving tricks as heuristics.

Not surprisingly, a brain built to make quick judgments frequently makes errors. Fast work is sloppy work. In that light, it would not really be fair to consider many of the mistakes that our brains make design flaws, seeing how well they perform in most instances. Limits, after all, are not the same things as flaws.

What *does* make cognitive biases qualify as defects is that they are not the result of an overtaxed system; they are patterns of mistakes that are made over and over again. Even worse, they are deeply ingrained and resistant to correction. Even when people know that their brains tend to get something wrong, and even when they're given all the information needed to get things right, there are some mistakes that they'll just keep on making.

For example, all of us are frequently guilty of something called confirmation bias. This is the very human tendency to interpret information in a way that confirms what you already believe is true rather than making a fair and objective assessment. Confirmation bias can take many forms, from selective memory to errors in inductive reasoning to outright refusal to acknowledge contradictory evidence. All of these are information-processing glitches that people usually cannot see in themselves even when they are pointed out but that they find immensely frustrating in others.

Most people's views on political and social policies are highly resistant to change, regardless of the data they are presented with. In a classic example, social scientists assembled a random group of individuals and showed them two (made-up) research studies, one seemingly proving that the death penalty was an effective deterrent to violent crime and the other seemingly proving that it was not. The researchers then asked the subjects to rate the quality and relevance of each study. Overall, the participants tended to give high ratings to the study that supported their own views and low ratings to the study supporting the opposing view. They would even sometimes specifically cite limitations of the opposing study that were also present in the study they agreed with! In other experiments, scientists have gone even further by giving participants fabricated studies regarding affirmative action and gun control, two political hot topics. These studies were more comprehensive and more powerful and gave clearer results than any true study that had been done. It made no difference. People rated a study as well designed if, and only if, it supported their views on the subject. (This research reflects another fact about confirmation bias: it pervades our political climate, which is why no one has ever changed his or her mind on an issue because of an argument on Facebook.)

Another manifestation of confirmation bias is something called the Forer effect, named for Bertram Forer, who performed a now-famous demonstration on a group of unsuspecting college students. Professor Forer asked his students to take a very long and involved personality test, a diagnostic interest inventory, and told them that he would use the results to create a complete description of their personalities. A week later, he provided each of them with a supposedly tailored vignette describing his or her personality in a series of statements. Here is what one of the students received:

1) You have a great need for other people to like and admire you. 2) You have a tendency to be critical of yourself. 3) You have a great deal

of unused capacity, which you have not turned to your advantage. 4) While you have some personality weaknesses, you are generally able to compensate for them. 5) Your sexual adjustment has presented problems for you. 6) Disciplined and self-controlled outside, you tend to be worrisome and insecure inside. 7) At times you have serious doubts as to whether you have made the right decision or done the right thing. 8) You prefer a certain amount of change and variety and become dissatisfied when hemmed in by restrictions and limitations. 9) You pride yourself as an independent thinker and do not accept others' statements without satisfactory proof. 10) You have found it unwise to be too frank in revealing yourself to others. 11) At times you are extroverted, affable, sociable, while at other times you are introverted, wary, reserved. 12) Some of your aspirations tend to be pretty unrealistic. 13) Security is one of your major goals in life.

Here's the thing: All of the students received the same personality description, though they didn't know it. Their ignorance was key to the experiment. Once they all received their "personal" and "tailored" personality descriptions, they were asked to rate its accuracy on a scale of 1 to 5. The average score was 4.26. If you're like me, you probably thought the report above described you pretty accurately too. And it does. It's pretty accurate for everyone because the statements are either so vague or so universal that each one can apply to almost anyone who isn't a total psychopath. "Security is one of your major goals in life." Who *wouldn't* agree with that?

When you read the statements thinking they are tailored for you, you don't critically evaluate what they are really saying (or not saying). Instead, the statements seem to confirm what you already think about yourself. Of course, if the students had been told they were reading just a random list of personality traits, they would probably note that some didn't really apply, but because they were told that the statements were written especially for them, they believed what they read.

This error in our information-processing ability can get us into very real trouble. Astrologers, fortunetellers, mediums, psychics, and the like are well versed in the finer points of the Forer effect. With a little practice, a huckster can use only vague hints from his mark to weave an elaborate tale that seems eerily accurate and applicable. The key is that the hapless victim has to *want* to believe what he's being told. For this reason, the Forer effect is often called by another name: the Barnum effect, after P. T. Barnum, who famously said, "There's a sucker born every minute." Considering how universal the confirmation bias is, Barnum's quip is a gross underestimate. At the current global birthrate, there are *two hundred and fifty* suckers born every minute—roughly one every quarter second.

## Let's Make a Memory

Like the human brain's ability to think logically, its incredible capacity for memory is a marvel. From the world capitals you memorized in seventh grade to the phone number of your elementary school best friend to your vivid recollections of trips, movies, and emotional experiences, there are literally *billions* of bits of information bouncing around in your head. Yet here, too, an amazing human feature is filled with bugs.

There are all kinds of flaws in the way our brains form, store, and access memories. For instance, most people have had the experience of recalling and enjoying a vivid memory for years only to find out later, through a recording or by comparing notes with others, that the recollection has major inaccuracies. Sometimes, people remember an event as a first-person experience when they were actually bystanders. Other times, they translocate the memories to a different time or place, or they change the cast of players involved.

While these small errors may seem innocuous, they can have big consequences. To find them, look no further than the world of criminal justice.

If a prosecutor has an eyewitness to a crime, a conviction is usually a slam dunk. If a witness positively identifies someone as the assailant whom he saw commit the act, how could he be wrong about that? If the witness had never met either the accused or the victim before, why would he lie?

But researchers in the field of forensic psychology have made startling discoveries regarding the reliability of eyewitness testimony. Although you'd never know it from how police and prosecutors pursue and present evidence, there is at least three decades of research proving that eyewitness identifications are extremely biased and often mistaken, especially when it comes to violent crimes.

Psychologists have used simulations to show how easily memories can be distorted after the fact, and these shed light on what is going wrong in the brains of many eyewitnesses. For example, researchers recruited volunteers and randomly assigned them to two groups. Both groups watched a video of a simulated violent crime from a fixed and limited perspective, as though they were bystanders to that crime. Afterward, both groups were asked to give a physical description of the assailant. One group was then left alone for an hour while the other group's members were shown a lineup of possible assailants and asked if they could identify the perpetrator. However, there was a slight trick with the lineup. None of the actors was the perpetrator, but one — and only one — matched the rough physical description from each eyewitness in terms of height, build, and race. More often than not, the witness identified that person as the perpetrator, and in a majority of those cases, the witness was "very sure" that the identification was correct.

That's a troubling outcome, of course — but it's not the most disturbing part of this experiment. Sometime later, both groups were again asked to describe the perpetrator of the crime. The members of the group that had not seen a lineup gave pretty much the same descriptions that they had before. However, most of the people in the other group gave much more detailed descriptions. Seeing the lineup somehow "improved" their memory of the perpetrator. The increased detail that they provided always matched the actor *in the lineup,* not the actual perpetrator of the crime they had witnessed. When the researchers pressed the witnesses about their memories of the crime, they discovered that they were honestly reporting their recollections to the best of their ability. Their memories had been warped.

This work has been expanded in lots of interesting ways and has affected how lineups are done in most states. Experts in eyewitness memory tell us that the only valid way to do a lineup is to have every single person in it—the suspect as well as the foils (as the paid lineup actors are known)—match every part of the physical description that is given by the witness. And if the eyewitness description doesn't match a suspect perfectly (which happens a lot!), the foils must match the suspect, not the description. Moreover, conspicuous identifying marks and even clothing must be as similar as possible among the people in the lineup. Scars and tattoos should be covered, because if the witness remembers that the perpetrator had a neck tattoo and only one of the men in the lineup has a neck tattoo, there is a good chance she will identify that person, even if he is innocent. Her memory of the crime will then be retroactively edited with the new person's face spliced in. Even clothing can trip this memory-editing property of the brain, and it all happens without a person's conscious awareness. The false memory is as vivid as a real one. More, actually!

Bystander memory is bad enough, but memories of one's own experiences are even worse. For instance, it turns out that personal traumas are also vulnerable to memory distortion. This has been documented in cases of a single traumatic event, such as a sexual assault, and sustained stresses that might involve multiple types of trauma, such as being in a war. The memory distortion most often observed with regard to trauma is that people tend to remember suffering more trauma than they really felt. This usually translates into greater severity of posttraumatic stress disorder (PTSD) symptoms over time as the remembered trauma grows.

Not surprisingly, this deepens and prolongs the suffering associated with the trauma. In one example, researchers asked Desert Storm veterans about certain traumatic experiences (running from sniper fire, sitting with a dying solider, and so forth) at one month and then at two months following their return from service. Eighty-eight percent of veterans changed their response to at least one event, and 61 percent changed more than one. Importantly, the majority of those changes were from "No, that did not happen to me" to "Yes, that happened to me." This overremembering was associated with an increase in PTSD symptoms.

Researchers led by my colleague Professor Deryn Strange at John Jay College demonstrated this memory distortion with a clever set of experiments. They asked volunteers to watch a short film depicting a real fatal car accident in graphic detail. The film was split into a series of scenes separated by blank footage. Those blank spots represented missing elements — scenes that had been deleted. Some of those missing scenes were traumatic (for example, a child screaming for her parents), while others were nontraumatic (for example, the arrival of a rescue helicopter). Twenty-four hours later, the viewers returned and were given a surprise test probing their memory of the film they were shown as well as their thoughts and recollections about the film.

The participants scored well on their ability to recognize scenes that they had indeed been shown. However, about one-fourth of the time, they "recognized" scenes that they hadn't actually viewed. They were far more likely to overremember the traumatic scenes than the nontraumatic ones, and they did so with confidence.

In addition, some viewers reported symptoms that were analogous to PTSD. They reported thinking about the traumatic scenes when they didn't intend or want to and avoiding things that reminded them of the film. Interestingly, those with the PTSD-like symptoms were more likely than the others to overremember traumatic elements of the film that they hadn't actually seen. This is further evidence of a link between PTSD symptoms and memory distortion.

This consistent quirk of memory formation demands an explanation. Why would a brain with such exquisite cognitive abilities engage in self-harm by exaggerating past trauma? Is this simply an error and nothing more? Does the human brain, which after all only recently evolved such complex cognitive functioning, get overwhelmed in times of great emotional stress and make sloppy mistakes?

Maybe, but there might be a more interesting explanation. This process of false-memory formation could actually be adaptive. One possible benefit of exaggerated trauma recall is that it might serve to reinforce fear of a dangerous situation. Fear is a powerful motivator and a very important conditioning mechanism for avoiding danger. Normally, fear and aversion toward something eventually wanes if people are not repeatedly exposed to it. The strange quirk of remembering traumatic events as even more traumatic over time may serve to counter the normal tendency of fear to wear off. So again, we have a feature that is a bug—or vice versa.

## The House Always Wins

As bad as humans are at accurately remembering things that happened in the past, they may be even worse at evaluating things experienced in the present. This is pretty remarkable when you consider how essential this basic skill is to our species' survival and well-being.

As you go through life, you are constantly bombarded by input from the world around you, and you can navigate this sensory storm only by making countless, often very quick decisions, hopefully more good ones than bad. To do that, you have to assign value to various things, people, ideas, and outcomes. Your brain then measures the value of various outcomes and makes decisions that add or preserve value rather than deplete it.

Psychologists and economists agree that the way that people behave at a gambling table is just about the purest manifestation of what can go wrong in how humans measure value, particularly the value of money. Most humans are not very good with money. Because money can be gained and lost so quickly and easily, gambling is an incredibly focused way to probe for deep truths about how we approach problems of valuation. A great deal of psychological and economic research has focused on how humans make choices while gambling.

This is not just an academic exercise. Gambling behaviors translate into many other areas. Depressingly, the lessons researchers learn about decision-making at the gambling table are often generalizable to how people live their lives.

For starters, most people are terrible at the basic logic of gambling. Of course, the whole enterprise of gambling is illogical to begin with; the odds always favor the house. People know this. They know that casinos make money at their expense. Yet they gamble anyway, presumably because the thrill of the experience is worth something to them. They *value* the experience of it and consider gambling a hobby like any other—golfing, going to the movies, whatever. The money lost at the gambling tables is like the price of admission—no big deal. They know this at the outset and enjoy the exhilaration that comes with throwing their hats into the ring for a big payoff.

But gambling is different from other forms of entertainment in one crucial way: people dramatically, and consistently, overpay for it. The vast majority of people who go to casinos walk out of them having lost more money than they intended to. When you ask gamblers at the beginning of the night how much they are prepared to lose and then ask them at the end of the night how much they have lost, more often than not, they have lost more than their set limit. In fact, if you told them at the beginning of the night the actual amount of money they would lose, most of them wouldn't enter the casino. It may be true that people enjoy playing at the tables, but the explanation that is often given after losing money — that it was really just about having fun — is an ex post facto justification that allows them to deny, even to themselves, that they have made poor choices.

The poor decisions that people make when gambling demonstrate some defects in the human psyche. Perhaps the most revealing ones — and the ones most worth mentioning — are those with parallels to other aspects of daily life.

Many gamblers begin the night with a set amount of money that they are okay with losing. Let's say that, for one gambler, it's a hundred dollars. When he sits at a blackjack table that has a five-dollar minimum bet, he will usually bet one or two chips. He wins some; he loses some. An odd thing often happens if he starts winning money: he bets more. This is the most illogical thing anyone could do. If you find yourself up fifty dollars and then start betting twenty dollars instead of your usual five, it will take you just two or three bad hands to lose what it took you ten hands to win. Remember, if you play long enough, the house will eventually win. By increasing your betting when you're up, you accelerate the pace at which you'll be giving your winnings back to the house.

If you find yourself up by a fair amount, you should celebrate your good luck by betting *less,* not more. If you do that, you have a fighting chance of going home without losing. Of course, the only real way to win is simply to quit altogether when you're ahead, but almost no one can do that. People who are that powerfully governed by logic are unlikely to find themselves in the casino in the first place.

Casinos know this well. What do they do when someone has a hot streak at the table? Why, they send him free drinks. If he keeps winning, he'll get a voucher for the premium buffet. If his good luck just doesn't seem to run out, he'll find himself with a complimentary room in the hotel. The more money he wins, the nicer the room. Casinos want their high rollers in deluxe suites designed to make them feel very powerful and important.

Why do the casinos lavish such gaudy gifts on someone who has just taken piles of their money? To ensure that he doesn't leave. The more gifts they give, the longer the gambler will stay. The longer he stays, the more likely he is to cough up his winnings. In fact, because of the false sense of his own skill he acquired while racking up his temporary purse, he'll probably end up losing far more than he would have tolerated had he not found himself up in the first place.

However cautious and determined people are when they begin, their good judgment goes out the window once they start to win. It's almost like they are determined to give back their winnings—and that is precisely what they do.

This behavioral defect can be seen at work in our daily lives. People become less careful when they have more resources, which only ensures that they will soon part ways with those resources. We all know people who are perpetually penniless, often for valid reasons, like being a student, having a low-paying job, or having financially burdensome family and living expenses. But when these paupers come into a small sum of money, what do they do? Too often, they blow it rapidly.

Why do they do this? They finally have some financial resources that could be used to pay off nagging debts, upgrade a car or apartment, or make some durable purchase or sensible investment, but no. They blow it on fancy clothes, expensive dinners, or nights of debauchery. This is not rational behavior. The joy they get from these dramatic expenditures is fleeting, while the debts they accrue are lasting. Most of us are very good at stretching a dollar when we have to but not so good at making frugal choices without pressure. A small windfall could be used to make wise purchases that would provide long-term payoffs and even help save money in other ways, but most people are not capable of making good choices in that scenario.

An even more widespread psychological flaw that is especially apparent in casinos is something known as the gambler's fallacy. This is the belief that a random event is more likely to happen if it hasn't happened in a while or that a random event that has just taken place is unlikely to take place again any time soon. Assuming that the events are unlinked, this is a complete delusion. In gambling, as in many situations in life, the past has no bearing on the present.

When I'm at a casino (and I do go occasionally, since—like everyone else—I'm not a perfectly rational being), one of my favorite activities is watching players at the roulette wheel. If the ball comes down on, say, 00 on the roulette table on one spin, there is nothing making it less likely to come down on 00 on the very next spin. The odds are exactly the same on each spin. Conversely, if a number has *not* been landed upon for several spins, the odds of it coming up in future turns are no greater than they were in the past. This is basic logic—yet, inevitably, you will see players who have successfully bet on 00 shy away from that number in the next few spins. Or, if this same number *hasn't* won in a very long time, you might see a bettor putting lots of money on it spin after spin. When it eventually wins, he will immediately start looking for another number to play—again, one that hasn't won in a while. Casinos are happy to list all the previous winning numbers on the roulette table. They know that it doesn't actually matter but that hapless gamblers like this hypothetical sucker *think* it does.

Why do people fall for these tricks? Do they believe that the ball or the wheel somehow *knows* what happened in the previous spin and that this somehow forces a different outcome on the next spin? Of course people don't consciously think this. But they do somehow think that the universe makes more sense than pure randomness would generate. Indeed, the gambler's fallacy is deep-seated in the human psyche and sometimes disguises itself as intuition. If someone gives birth to three baby girls in a row, many people are convinced that the next one will be a boy. If this happens, their instincts are validated. If it doesn't, they shriek, "Wow, *another* girl! What are the odds?" About 50 percent, actually. The three hundred and fifty million sperm that raced toward the egg that grew into that baby were blissfully unaware that three females had previously passed through there. Each birth is like a coin flip. A coin has no knowledge of its previous flips. It's possible to flip ten heads in a row. When you go to flip the eleventh time, the odds are fifty-fifty that it will be heads again.

What explains the gambler's fallacy? Evolution. Our brains are like computers that evolved mostly to run programs called heuristics. Heuristics are rules that the brain establishes in order to make sense of the world quickly in ways that will help it (hopefully) make good decisions. When you observe something, you often unconsciously translate it into a larger pattern and assume that what you have observed is representative of a larger truth. It's true that this skill was and is extremely useful. If an ancestor of ours observed a lion hiding in shrubby underbrush, she might reason that shrubby underbrush was a spot where lions could be found, and she would be careful to avoid areas like that in the future. She extrapolated a larger truth from a single data point—and possibly saved her own life in the process.

As useful as heuristics are, mental shortcuts can actually trip us up when we encounter a data set that is boundless. This is because the human brain is not built to comprehend infinity; we are trapped by the limits of finite math. For instance, when it comes to coin flips, we know that things *should* work out to fifty-fifty odds. So if someone observes four heads in a row, his brain applies this observation to a finite data set. The unconscious reasoning is something like *There were just four heads in a row, so there needs to be some tails coming soon in order to achieve the required fifty-fifty ratio.* This small-number thinking probably served our ancestors well in the development of pattern recognition and learning, but in modern times, it misfires in various ways, particularly when we're faced with probability and the math of big numbers.

Back to our gamblers. As if failing to quit while they're ahead weren't bad enough, people also have difficulties in quitting when they're already deep in the hole. How many times have you heard someone say (or said yourself), "Oh, just one more hand and I'll get it back," or the even more explicitly wrong "The house *owes* me after those last few hands," as if some ledger is being kept and the cards (or dice or roulette balls) must act to balance out past injuries. Nothing could be farther from the truth. When you're in the middle of a losing streak, it's worth remembering that the odds are slightly better that the streak will continue than that it will improve because, as always, the odds always favor the house.

The failure to quit when one is down is due to a related, possibly connected, fallacy termed *sunk costs*. Part of the reason that people have trouble walking away after losing money at the blackjack table is the notion that the money lost would be "wasted" if they didn't stay to try to win it back. Of course, this is a fallacy on top of a fallacy because nothing done in one hand can improve the odds for the next, but that doesn't stop people from thinking this way. The sunk-costs fallacy often gets wrapped up and packaged with wise legitimate investment practices, such as the notion that you have to spend money to make money and other such truisms regarding future rewards.

Remember: Not all money spent is an investment. Some money is simply lost, and trying to get it back should never be used as a reason to stay in a losing situation. If the dealer draws blackjack, neither the house nor the universe owes you a damn thing. This does not mean you are more likely to win the next one. You are in exactly the same place you were, just a little bit poorer. If the dealer draws twenty-one ten times in a row, he is just as likely to draw it again in the future. None of your lost bets buy you future odds. That money is simply gone.

The fallacy of sunk costs can be seen in every sphere of human activity, not just in the casino. For instance, many amateur investors— which is almost everyone with a 401(k)—will consider how much they spent on a stock before deciding whether or not to sell. This makes no sense whatsoever. The only factor you should consider in deciding whether to sell or hold a stock is your belief about its future performance. It doesn't matter if you purchased it a day, month, year, or decade ago. If you think the stock will go up over the time you can hold it, you should hold it. If you think it will go down, you should sell it. It's that simple.

Now, there might be good reasons to hold a losing stock. The price may be artificially low because of unfounded investor panic about the company or because of a temporary market depression that is likely to subside. Those are valid reasons. How much you paid for the stock when you first bought it is irrelevant. And yet, this is what people often consider most. In fact, most portfolio managing programs make this easy: They include a column, usually right next to the current value, indicating what was paid for the stock. This is a shame because it reinforces the notion that it matters how much you've lost or gained in the past in deciding what to do going forward. If a stock has been steadily declining, that's a sign that it may be time to sell. However, a lot of times, people put off the inevitable decision to sell to see if they can catch it on an upswing and at least get back what they paid for it. While they wait for that, the price continues to fall, and they lose more money.

It's not just the stock market. The fallacy of sunk costs can influence lots of financial decisions, usually for the worse. For example, when it's time to sell property, people are very reluctant to do so at a loss. They will hold on to houses and other property for extended periods, waiting for the market to recover so that they can recoup what they paid for it. That may sound like a sound financial move, but it costs money to hold on to properties; there are annual tax bills, utilities, and maintenance costs. People rarely consider those costs when they hold on to a house longer than they should. In addition, if the property is not being used for housing or to generate income, that capital is simply tied up doing nothing when it could be doing something.

The sunk-costs fallacy dooms decisions people make as individuals as well as ones they make as a society. Soon after the U.S. invasion of Iraq, it became clear that the continued military occupation of the country was no longer fruitful for anyone involved. The U.S. military had "won" the war by deposing and disarming the previous regime, but in the wake of this destabilization, widespread violence, terrorism, and chaos reigned. The United States continued its occupation with the goal of rooting out the insurgent fighters and bringing stability to the country. Eventually, the continued presence of American troops was itself a destabilizing force, the linchpin of radicalization and terrorist recruitment. Even after everyone began to acknowledge this grim reality, there was strong resistance to pulling military forces out of Iraq. The political arguments routinely invoked the "lives lost" and the "money spent" arguments. *We've already given so much—it can't all be for nothing!* America may certainly have a moral responsibility to try to help the Iraqi people, but that's a different issue—one whose solution cannot be a military one.

The sunk-costs fallacy pops up any time people feel that they've invested time, effort, or money in something and don't want to see that sacrifice go to waste. Of course, that is understandable, but it can fly in the face of logic. On occasion, it simply doesn't matter how much you've invested; sticking with a failing plan will only cost you more. In these instances, it can be very hard to see through your own stubbornness, but cutting your losses might be the smart thing to do.

## The Price Is Wrong

The gambler's fallacy and the sunk-costs fallacy are two specific ways that we screw up our lives when it comes to money or other resources, but it turns out that we make even more fundamental errors when it comes to things of value: we routinely goof up the process of assigning value in the first place.

Consider the games that retailers can play with price tags—and how effective their ploys are. For example, many studies have shown that consumers gravitate toward items that are marked as discounted, regardless of the actual final price. A twenty-dollar shirt will move much faster if it is priced at forty dollars and then discounted to 50 percent off. We humans measure value in relative terms, not absolute ones.

We also have an anchoring bias. People give a great deal of value to the first piece of information they receive, regardless of its trustworthiness. This leads all further information to be valued not in strict terms but relative to the original. Using the example above, the first piece of information is the original (inflated) price of the shirt. This makes the sale price of twenty dollars seem much lower by comparison.

The same is true in a salary negotiation or a home purchase. The first person to name a figure always sets the bar, and all of the parties in the negotiation perceive—and value—every counteroffer that follows relative to that opening bid. Savvy salary negotiators always make their first request way above what they think they will get; they know that this allows managers to feel like they got a "deal" by getting the employee to accept 5 percent to 10 percent less, even if this is still more than they initially intended to pay.

This cognitive bias is so ingrained in the human social psyche that people rarely even question it. I experienced this firsthand when I was contacting solar-power companies to come look at my home and make proposals—I found myself comparing every offer with the first one. The first offer happened to be way inflated because the company did not have a history of building the kind of system that I wanted and basically didn't want the job. After getting several lower bids, I was left feeling that installing solar panels was cheap! Only when I discussed the bids with my dear spouse was I reminded that these offers were all way above what I originally said I would spend.

Why did the people at that first company give me a sky-high bid instead of just passing on the job? Perhaps they felt that if they charged me way too much, it would make up for however much it cost them to take on a job that wasn't really in their wheelhouse. More likely, however, they knew that by giving a very high bid, they would leave me with the impression that they were the Cadillac of solar companies. It worked too! Several weeks later, I caught myself telling a friend, "Now, if you can afford it, probably the highest-quality company out there is . . ." What was I talking about? I had no idea about the quality of the craftsmanship of that or any other company. All I knew was the *bid*, but that was enough. The contractors had successfully convinced me that they ran a superior company by inflating their price, and I was perfectly happy to serve as their unpaid spokesman.

Valuation biases are well known to marketing and sales professionals. The beverage industry is just one of the many sectors of the economy where these professionals can draw on scientific research to sell more products. For instance, studies have shown that people will avoid wine that is priced too low because they think it reflects poor taste or quality. Blind taste tests have demonstrated that people's perceptions are actually altered when fake price tags are put on bottles. Faked higher prices make people like the wine better, while lower price tags (but not actually cheaper wine) make people turn up their noses. When the deceit is revealed, the participants in these studies, while often embarrassed, frequently confess that the higher-priced wine really did seem to taste better. It was not merely a matter of trying to impress the researcher; false valuation actually affects perceptions, including taste.

As wine salespeople know, valuation biases can work in the opposite direction as well. Next time you go to a nice wine store, take note of the prices. A single bottle of expensive wine is often thrown in with middle-priced bottles to make those bottles look cheap by comparison. That single bottle might not actually cost that much. It could be a very cheap bottle with a big inflated price tag. It's just there for show anyway!

Similarly, a bottle of cheaply priced wine can be thrown in to make the others look more respectable. This, too, can be done with false pricing. When a wine merchant is down to her last bottle of something, she'll often mark it way down and then use it to help move some other kind of wine that isn't selling well. Let's say there are ten-dollar bottles of Merlot that haven't been moving. If she sets a bottle with a six-dollar price tag right next to them, suddenly they start to look better. Of course, once she's ready to unload the six-dollar bottle, all she has to do is put a fifteen-dollar price tag above the other one and then put a big *X* through it. It'll sell in minutes.

By now, you have probably realized that many of the cognitive biases and errors common in humans manifest most acutely in the handling of money, whether it's gambling, markets, or financial planning. Of course, currency is a human invention with no direct correlate in the natural world, and for most of human history, economics involved the exchange of actual commodities — things that were useful in and of themselves, not arbitrary representations of value. Thus, it shouldn't be surprising that we haven't evolved the cognitive skills to manage currency. It's purely a conceptual construct with no biological basis, and that's why so many people choose to rent their homes and buy their cars instead of the other way around.

But while money itself is relatively new, the ways in which we misuse it reflect ancient glitches in our species' mental circuitry. That claim is less surprising when you consider that, although human cognitive capacity evolved in a world where currency did not exist, resources certainly *did* exist, and thus so did the concept of value and its impact on decisions. Humans have always had a relationship with goods, services, and real estate, the tangible things that bring value to the possessor. Goods can be things like food, tools, or even trinkets. Services can be things like cooperation, alliances, grooming, midwifery (yes, it has existed for a long time!), and so on. Real estate means that certain locations are more coveted than others for building a camp, nest, hunting blind, and so on. Economic forces, in other words, predate currency by quite a long time.

Although it's hard to compare our species' current relationship with valuable resources to its past one, insofar as we can measure it, other animals make many of the same errors that we do. For example, many animals purchase sexual access with food or other gifts. There are penguins that trade sex for nest-building materials. (If you're interested, I have a whole section on prostitution among animals in my book *Not So Different*.) In bird colonies, nest location often correlates with social rank, and there is the equivalent of a bustling real estate market complete with attempted evictions and robberies. Nature offers many examples of how animals try to dominate resources above and beyond what they need to thrive and reproduce. Greed and envy are not uniquely human traits. Our species may have invented currency, but we are not the first to engage in economic transactions, and thus neither are we the first to confront issues of economic psychology.

Just how similar our flawed economic thinking is to that of other animals is becoming clear, thanks to research involving our primate cousins. Dr. Laurie Santos, an animal behaviorist and evolutionary psychologist, has spent years establishing "monkeynomics," an environment in which capuchin monkeys have been trained to use and understand currency. From the many papers she has published on this fascinating work, the most important lesson is that monkeys exhibit many of the same irrational behaviors as humans do when it comes to resources. They are loss-averse, meaning they will take foolish risks when faced with the prospect of losing "money" that they've already earned, risks they wouldn't take in order to earn the same amount. They also, like us, measure value in purely relative terms and can be tricked into changing their choices by price manipulation, like we are at the wine store.

The fact that monkeys have a lot of the same cognitive flaws that we do points to a deeper evolutionary truth about our defective economic psychology. Behaviors that we now see as erroneous and irrational — falling for one's own confirmation bias, for instance, or making decisions based on the sunk-costs fallacy — likely served our preagriculture ancestors well, as such things as roulette tables and beachside condominiums had not yet been invented. Likewise, when resources were used purely for subsistence rather than social status, comfort, or power, the system of measuring value in purely relative terms probably made good sense.

Furthermore, the stakes — and thus the evolutionary pressures — are high for animals in the wild, and our ancestors were no different. For most modern humans in the developed world now, losing some money generally means that they might have to scale back some aspects of their lifestyles. Losing resources in the Pleistocene epoch might have meant starvation. Thus, extreme aversion to loss also made good sense. When the alternative is an almost certain death, taking a risk doesn't seem so foolish. Desperate times call for desperate measures.

So the flaws in our economic thinking did serve an evolutionary purpose. But as wine merchants, the casino industry, and many other economic opportunists know well, this feature is a significant bug.

## Power of the Anecdote

Another form of human irrationality is found in our species' extreme sensitivity to the influence of anecdotes. Often, one particular event in your life, or even something told to you by someone else, can over-power everything else you know about the phenomenon in question. This effect is a subset of a larger fallacy known as neglect of probability.

I was once a passenger in a friend's car as he prepared to merge onto an interstate highway from the city streets. As he approached the point of merger, he slowed the car down and eventually came to a full stop so that he could look over his shoulder and observe the oncoming traffic on the highway behind us. I shouted in disbelief, "What are you doing?" He responded, "One time, I was merging into highway traffic and got into an accident, and now I just wait until there is a total break in traffic before I get on the highway."

My friend had succumbed to the power of an anecdote. Both driver's education courses and the rules of the road agree that it is safer and more efficient to keep moving as one merges into traffic from a ramp. There is ample road to accomplish this on interstate highways. It's actually *dangerous* to stop because cars could be approaching at high speed from the ramp behind you and, depending on road and lighting conditions, may not be able to come to a full stop in time to avoid a collision. My friend had no doubt been in cars many times when drivers, including himself, safely merged into highway traffic. Yet one bad incident totally changed his thinking and his behavior, leading him to drive *less* safely in an attempt to drive *more* safely.

Of course, large data sets are simply collections of individual anecdotes — but their size is what makes them so powerful. By assembling and analyzing lots of different data points, researchers can find statistical patterns and hidden truths that individuals relying exclusively on their own limited reservoirs of experience cannot possibly see. Yet anecdotes convince us when statistics cannot, because data doesn't move us; stories do. Stories carry more weight with us than generalized statistics do because we can relate to the protagonists of a story and feel empathy for them. We cannot feel empathy for data.

Playing the lottery is a different form of worshipping anecdotes over statistics. My parents have been playing the lottery for as long as I can recall. They don't waste their money on scratch-offs and small payouts in the pick-three and pick-four games. No, they go for the big lotteries, which promise life-transforming amounts of money. Over the years, my parents have surely thrown away tens of thousands of dollars that could have been put to very good use, as my parents are penny pinchers who use their money wisely in every other way. Any time I remind them of this, my mother defends herself with the "purchasing hopes and dreams" logic that is as flimsy as it is common. After all, hopes and dreams are free.

My parents, like everyone else who plays the lottery, are moved by stories of nurses winning a million dollars. They see people on television receiving the check and think, *That could be me!* What they don't see is the several million people who are each a few dollars poorer for having purchased the lottery ticket. The power of the anecdote reigns supreme.

People often combine the power of anecdotes and the confirmation bias to support their positions on any number of social issues. If you think government welfare is wasteful, you probably have a ready example to prove your point. If you think that corporations ignore environmental destruction, you likely can list catastrophes wrought by evil industries. You surely can recount exactly why so-and-so is the best quarterback in the NFL. Although none of these have as much evidentiary power as a large set of data and accompanying statistical analysis, they're much more convincing in an argument. And that's just bananas.

The reason that anecdotes are so much more powerful than data is once again that our minds are trapped in the world of finite math and small numbers. Our brains evolved on a planet where humans came into contact with no more than a couple hundred people in their lives. It was crucial for them to draw conclusions from what they saw and what they learned from others so they wouldn't have to learn every lesson themselves. Questions about how government policies affect millions of people were just never on the docket while our species was taking shape. Today we can use pen and paper (or, um, computers) to crunch the numbers, but our brains will never really be able to wrap themselves around large numbers, even when we're able to complete math problems mentally. You might be able to compute ten million times three hundred billion in your head, but you can't really *comprehend* ten million of anything.

Because early human societies were never larger than a couple hundred individuals, there was no need to understand mathematical conceptualizations larger than that. Therefore, that skill simply didn't evolve. Some have even argued that the human brain is naturally wired to understand only three numbers: *one, two,* and *many,* an argument bolstered by the discovery that the Pirahã tribe of South America has words for only those three numbers. While there is vigorous debate on what numbers, if any, humans have hardwired conceptions of, there is little debate that the human brain is poorly designed for mathematics.

We often pay a price for that—literally, in the case of lottery addicts. But when it comes to another one of our cognitive defects, the cost is even higher.

## Youth Is Wasted on the Young

The clichés are well known. Old people drive slowly and carefully and always wear their seat belts. Young people are reckless and inattentive behind the wheel.

These statements seem so obvious that we often forget how paradoxical they are. Shouldn't young people, with their whole lives ahead of them, be more careful when operating the deadliest machine most of them will ever come in contact with? And shouldn't old people, with such precious little time left, want to get places quickly?

This conundrum is more than just a cliché, however; it's backed by data. Young people are indeed the most dangerous drivers, the least likely to buckle their seat belts, and largely indifferent to safety features when selecting cars to purchase. Youthful reckless driving is not a function of inexperience either, because studies have shown that novice drivers who are more mature when they begin to drive are almost as successful at avoiding accidents as more experienced drivers. It's age, not skill, that appears to lead to caution and care behind the wheel— something that car-rental companies have long recognized. They often refuse to rent to drivers under the age of twenty-five. If it were simply a matter of experience, the policy would instead be against drivers with fewer than eight or nine years of driving experience, but it's not. It is youth itself that makes some drivers just not worth the risk.

Of course, this phenomenon goes well beyond driving. Young people are bigger risk-takers in pretty much every way. They try dangerous and illegal drugs at much higher rates. They're much less likely to use protection when having sex. Young people are much more likely to enjoy extreme sports such as bungee jumping, skydiving, rock climbing, BASE jumping, and so on. Even when there are safety measures in place, these activities are pretty risky, and in any case, they give the people who participate in them the *perception* of danger, which is at least partly responsible for the thrill. Psychologists often treat habitual thrill-seekers as addicts—they're addicted to the adrenaline rush that accompanies danger.

Young people seem to actually *enjoy* being in danger. One manifestation of this phenomenon is cigarette smoking. Young people will take up this habit even though they all know cigarettes are deadly. In fact, smoking is very *unpleasant* the first time you do it. I remember my first few cigarettes. I felt a harsh, stinging pain in my throat, and I immediately coughed. The nicotine made me feel lightheaded and eventually nauseated. Some people even vomit while "enjoying" their first cigarettes. Despite the unpleasantness, I didn't stop. I actually *worked* to get myself addicted to smoking. Each cigarette was a little bit easier to take, and the nausea eventually gave way to a mild relaxation. By then, I was hooked, and it took me twenty years to kick the habit. Six years after finally quitting, I still curse myself for having started in the first place.

If someone makes it to the age of twenty-five without picking up a cigarette, he has an almost zero chance of starting after that. The label of nonsmoker is pretty firmly established even by the time a person hits twenty-one. As people mature, they become far too wise to start a habit as foolish as smoking. Especially since the first try is so awful—who the hell would be irrational enough to keep at it? Well, I was, and so are millions of other kids. So the question is not *who,* but *why?*

The key to understanding risky behaviors lies in another basic fact: They are more prevalent not only in young people, but also in males. Young adult males are the most dangerous demographic of the population, and the reason for that is even stupider than the risky behaviors themselves: they do stupid things in order to impress people.

Young people, particularly but not exclusively males, take insane risks in order to advertise their "fitness." This does not necessarily mean displaying physical prowess, although it can certainly be that. The term *fitness display* comes from the study of animal behavior and refers to ways that animals communicate to potential mates, as well as potential rivals, that they are forces to be reckoned with. It's a way of saying, *I am so powerful that I can do this dangerous thing and still come out okay.*

If we use the analogy of smoking, the coded message is *I am so healthy that I can survive doing what we all know is incredibly unhealthy.* If young men were taking these risks purely for their own thrill, they would take them when they were alone, but they don't (until, in the case of cigarettes, they become addicted to the nicotine). The crazy behavior is always done in full view. The larger the audience, the better.

Displays of fitness have a long evolutionary history, and there are many different types, but for our purposes, the ones that matter fall into a specific category that biologists call "costly signals." Costly signals are responsible for some of the more conspicuous examples of sexual selection found in nature. For instance, the enormous tail of the peacock and the bulky antlers of the stag serve no other purpose than attracting mates. And they are costly; it burns many calories to carry them, and they reduce the animals' speed and mobility. While it is true that some horned mammals use their horns to fight among themselves, many never or rarely do so. The antlers and the tails are mostly for display, to show the females how strong they are. How does a huge tail show off strength? Well, you try lugging that enormous thing around everywhere you go without starving to death or getting killed.

The handicap principle states that sexual selection sometimes leads to the evolution of ridiculous impediments to survival that exist purely so the male can display feats of strength. These examples of runaway selection are hardly good for the overall health and vitality of the species, yet those enormous antlers that are purely for show? They work! Female deer are strongly attracted to large and elaborate antlers. Size matters. Same for the tail of the peacock. Peahens dig it.

Applying the handicap principle to behavior is tricky, and applying it to humans is trickier still. However, there is solid evidence for doing so. Studies have confirmed that young females experience stronger sexual attraction to males who display risky behaviors, particularly those involving feats of physical strength. Young women will score the same male as more attractive when they see him playing football than when they see him playing the piano.

Even more telling is the fact that males are impressed by their fellow males when they see them taking risks. Doing things like drag racing, cliff diving, and smoking cigarettes helps a young man earn and keep male friends. In evolutionary terms, these are male-male alliances, and they can prove very valuable in securing a prized place in the dominance hierarchy of social species up to and including humans. And if you're a male, the higher you are in your dominance hierarchy, the greater your chances of reproductive success. We humans like to think we're above all that, but I suspect everything in this section sounds familiar to anyone who survived high school.

By contrast, young males tend to be more sexually attracted to females who take *fewer* risks. This may help explain why males are drawn to taking risks more than females are—there is a potential payoff for them but not for females. This also supports the notion that, in mammals, individual males are fairly expendable, while females are a limiting factor for the survival and success of the species. In this view, each female is precious, and males are generally drawn to females who show signs of caution and care and, thus, will be able to ensure the survival of their children. Females, however, worry less about caution in their mates and instead want good genes for their offspring.

This is a dramatically overgeneralized view, of course, but like the clichés about age and risk-aversion, it has some grounding in truth. It's basically a truism that the jocks get the girls in high school while the geeks get passed over, even though the latter are probably more likely to experience real-world success later in life. By that time, the tables get turned, but it's too late for many women and men because they're done reproducing. And while it might be tempting to think that the recent increase in the age of first reproduction among humans will reduce costly displays of fitness among young men (and make smart, sensitive young men more attractive to their peers), this is unlikely to have any quick impact. An evolutionary transformation like that would require a genetic difference between risk-takers and safe players, followed by many generations of selective pressure. Absent those factors, humans can count on adolescent boys doing stupid things for some time to come.

An important implication of this bug in our brains is that much of the public-awareness programming aimed at reducing rates of smoking, drinking, drug use, and other risky behaviors may be totally backward in its approach. While explaining the risks of drugs to a high school student might seem like a logical way to dissuade him from trying them, it probably has the opposite effect. Explaining that drugs are risky could make them *more* attractive to kids, especially young boys. There is wisdom, indeed, behind another old saw: There is no better way to raise interest in something than by banning it. In the primate brains of young boys, if drugs are dangerous and illegal, the people who try them must be really strong and brave. Could there be a clearer cognitive flaw than that?

## Coda: Saints and Sinners

How humans became so much more intelligent than our closest relatives in such a short time is one of the biggest mysteries of evolution. While it's obvious that high intelligence has survival advantages and would thus be favored by natural selection, it is actually a fairly improbable feature for a species to evolve.

First of all, evolving to be smarter requires a fairly ordered progression of a lot of mutations, such as those allowing the expansion of the cranium, the growth of the brain itself, greater interconnectedness of brain areas, and so on. Second, at least in mammals, it also requires changes in female reproductive anatomy to accommodate larger craniums during childbirth. Third, brains are extremely energy-hungry and thus place intense demands on organisms to acquire enough calories to support them. For example, the human brain consumes about 20 percent of the body's daily energy expenditures, more than any other single organ. The fact that so many long-persisting lineages, such as sharks, horseshoe crabs, and turtles, never bothered to evolve big brains underscores how costly and improbable it is.

But big and smart brains *did* evolve in humans, despite the costs and anatomical constraints. Therefore, this would seem to be a triumph of *good* design, hardly a defect. But a closer look reveals that our large and powerful brains may actually be the biggest flaw of all.

Most anthropologists agree that the initial phase of the expansion of human intelligence took place over the first four or five million years of our species' divergence from the chimp lineage and was marked by the shift to larger and more intricately cooperative social groups. As our ancestors shifted to a bipedal posture and eked out a living at the borderlands of dense rainforests and grassy savannas, they began to creatively master a broader range of survival skills. They needed expanded cognitive abilities to *perform* these complex skills but also to *learn* them. Through the human lineage, our species gradually made a transition from preprogrammed behaviors and skills to learned ones. Much of the learning took place socially, with one individual teaching another. Thus, skills and social interactions were linked and both evolved together, pushing the human brain toward ever-greater abilities.

As upright posture freed our ancestors' hands to carry things and

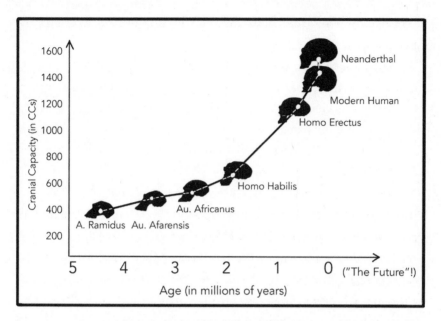

Our ancestors' cranial capacities expanded gradually for most of the past five million years but then accelerated in the last million and a half years. This dramatic acceleration may reflect the development of new, antisocial competitive strategies.

make tools, and as growing brains and larger groups facilitated social learning, humans found themselves in the perfect environment for the emergence of more complex forms of communication and cooperation. Cooperation requires perspective-taking and empathy; in order to really cooperate with you, I must be able to imagine how events will unfold from your point of view. In order for people to work effectively in teams, each member of a group must have some appreciation for what the other members in the group are seeing, thinking, and feeling. Our ancestors took cooperation and sociality to dramatic new heights, and their powerful intelligence played a key role in that. Until . . .

About a million and a half years ago, brain expansion in our ancestral lineage suddenly accelerated dramatically. Human brains have grown more than twice as much over the past one million years than they did in the five million years before that. What happened to facilitate that rapid change?

Recent studies have indicated that the rapid acceleration of brain growth in our ancestors was likely due to a switch to more competitive strategies for survival. At this time, there were several species of hominids vying for similar habitats and resources. In addition, even within the same species, different social groups would compete with one another when their territories overlapped.

Competition among groups of animals is nothing new, of course, but our ancestors approached this competition with dramatic new cognitive abilities. This is where things get very dark.

Humans engage in competitive behaviors that are downright Machiavellian. We manipulate, deceive, entrap, and terrorize. To do this, we use many of the same skills that serve us well for cooperation: perspective-taking, predicting another person's next move, and so on. In other words, throughout our evolutionary history, we began using our impressive cognitive powers for good, but then we turned to the dark side. And, like Anakin Skywalker becoming Darth Vader, that's when we grew *really* powerful.

To see the evolutionary legacy of this adaptation, look no further than today's headlines. Humans are capable of unspeakable violence toward one another. We plot against one another with ruthless cunning and complete disregard for others' suffering. The surprising part is that our ancestors didn't sacrifice their cooperative, pro-social, even altruistic nature in the process of becoming so ruthless. They kept both sides — becoming a species of Dr. Jekylls and Mr. Hydes.

The duality of human nature is the hallmark of the human experience. We can switch from boundless love and great self-sacrifice to cold-blooded murder, even genocide, in a heartbeat. Just a few generations ago, the United States of America and many other countries were populated by men who were loving fathers and doting husbands and who made their fortunes by brutally enslaving other human beings. By all accounts, Adolf Hitler was a generous and gentle partner to Eva Braun even as he ordered the senseless slaughter of millions.

How can such unspeakable monstrosity and genuine affection co-exist in the same species, let alone the same individual? Because evolution rewarded those of our ancestors who could nimbly switch between cooperation and competition when the conditions suited them. We evolved to be highly social, collaborative, and altruistic but also ruthless, calculating, and heartless. It is this latter set of traits that seemed to facilitate the evolution of our great big brains. So the next time you praise someone's intelligence, stop to think about what — or, rather, who — had to be sacrificed in order for her to get that smart.

# Epilogue:
## The Future of Humanity

*Why humans are still evolving, despite what you may have heard; why all civilizations, including our own, may be destined to implode and rebuild in an endless cycle; why we may live indefinitely healthy lives in the not-too-distant future; why technological advancement has both raised the possibility of our self-destruction and provided us with the means to avoid it; and more*

This book has surveyed only a fraction of the defects in the human body. We have many other mental biases, countless additional problems with our DNA, and lots of other useless (or needlessly complex or breakable) body parts among myriad sundry flaws not mentioned here. Any book attempting to cover all human imperfections would need to be much, much bigger than this one—and much more expensive. You're welcome.

That said, our many flaws should not make us feel bad about ourselves. After all, evolution works by random mutation and survival of the *fittest,* not of the perfect. Such a haphazard approach to life could *never* produce perfection. Every species is a balancing act of positives and negatives. Great as humans are, we are no exception.

But when it comes to imperfections, the human story *is* unique. We do seem to be more flawed than other animals, and the reasons why, paradoxically, come from adaptations that really should have been improvements. For example, the reason we must pursue such a diverse diet in order to stay healthy while other animals can survive on a single type of food is that our ancestors were able to break free from the monotony of subsisting on staples and use their great cognitive powers to forage, hunt, gather, dig, and otherwise seek nutrition from every conceivable food source across multiple habitats. That sounds like a *good* thing, but the problem is that, as their minds grew powerful, their bodies grew lazy. Basking in the variety of their rich diets, their bodies simply quit bothering to make many nutrients that they previously did. This forced our ancestors to switch from being able to *enjoy* a rich diet to *requiring* a rich diet in order to survive. That's an unfortunate switch. What was clearly an advantage to start with—being voracious omnivores—became a limitation.

This same logic applies broadly to human anatomy and physiology. Our species' physical form is the result of compromises forged by evolution as humans became the ultimate generalists. There are species that can run faster, climb higher, dig deeper, or hit harder, but humans are special because we can run, climb, dig, *and* hit. The phrase *jack of all trades, master of none* fits us perfectly. If life on earth were like the Olympic Games, the only event that humans would ever win is the decathlon. (Unless chess became an Olympic sport.)

Other problems we experience with our bodies are due to the many differences between the environment our ancestors evolved in and the environment humans now live in. Those differences lead to so-called mismatch diseases, such as obesity, atherosclerosis, type 2 diabetes, and many other disorders. Much of these problems of environmental mismatch stem from differences between our ancestors' diet and ours, but another big difference between how our ancestors lived in the early Stone Age and how we live now has to do with technology. Because technology has allowed us to move beyond the physical limitations of our bodies, it would seem like a purely advantageous phenomenon. However, the less we rely on our bodies, the less they are pressed to adapt and evolve. Now that we are solving so many of our problems with technology instead of biology, it's no surprise that our bodies are not exactly in tiptop shape.

Of course, humans are not the only species to use technology. For our purposes, I'm defining *technology* as methods, systems, or devices crafted for their utility in performing a task, and by this broad definition, many animals make use of technology. Macaques use rocks to crack nuts, and chimpanzees shape sticks to use in their hunt for termites, to give a couple of examples. In our species' case, early humans used simple stone tools. But unlike macaques and chimps, who are still using the same tools that they have wielded for millions of years, humans' invention of stone tools ushered in a new kind of evolution that has set us apart from every other animal under the sun and for which there is no turning back: cultural evolution.

Cultural evolution refers to the social practices, knowledge, and even languages that are passed on through the generations. While animals certainly learn some things from one another, humans have taken the concept of culture to the extreme. Almost everything we do and experience in our lives is the result of culture, and this has been the case for quite a long time. Once modern humans began sharpening rocks, building dwellings, and, eventually, planting crops, they began to succeed or fail based on cultural features rather than biological ones.

In a sense, we have taken charge of our own evolution, but are we really in control? As technology and culture continue to advance, what changes are in store for us? Now that we understand how our biology and culture evolved, are we able to manipulate them at will and therefore shape human destiny in a deliberate and intentional way? Or will we continue to plow ahead in the same random, haphazard manner that we have for the past seven million years? In short, what does the future hold for our species?

## Are We Done Evolving?

Some high-profile people in the scientific community, including Sir David Attenborough, have asserted that humans have developed so far in our civilization and technology that we have fully escaped the forces of evolution. We are no longer evolving, they say, and our species will remain more or less the same in its biology, absent any intentional tweaks that we accomplish.

There may be some truth to this. Existential challenges are the hallmark of evolutionary theory and one of the key observations that led Darwin to his great discovery. Yet today, we face very few of these challenges relative to the thousands of generations that came before us. The vast majority of human beings who are born today will survive to reproductive age. Starvation is rare, at least in developed nations. Physical injuries and illnesses are now routinely overcome by modern medicine. Fighting to the death is also rare. Murder is punished. Even warfare has waned tremendously. A nice long life is pretty much assured for most people alive today.

Furthermore, reproduction is not nearly as competitive as it once was. Although those with great physical strength and stamina may attract more desirable mates, they don't, as a rule, leave more offspring. The same is true for intelligence, a strong work ethic, or good looks. For humans through the Pleistocene epoch, things like eyesight, dexterity, speed, endurance, intellect, popularity, health and vitality, dominance status, and even attractiveness had a direct impact on the number and success of one's children. But nowadays, being successful at life, either socially or professionally, doesn't generally mean that someone leaves more offspring. As I'll explain in a moment, it may actually mean that he or she leaves less! Even major medical problems and limitations don't automatically reduce the risk of having successful children. The usual forces of natural selection have largely been neutralized.

Natural selection may not be shaping us anymore, but evolution is still at work. Evolution simply refers to any genetic change in a species over time. Natural selection, the phenomenon that picks winners and losers through their survival and reproduction, is just one way that a species can evolve. It's the one we think the most about, but there are other evolutionary forces that can be just as powerful. So, yes, humans may indeed have managed to escape the scourge of natural selection, but that doesn't necessarily mean that evolution is over for us.

A species can evolve any time that reproduction is nonrandom. If some specific group of individuals reproduces more than other groups, that group will contribute more to the gene pool of the next generation. Assuming that the difference inherent in that group has a genetic component to it, this demographic change makes evolution inevitable simply by introducing a gradual genetic change into the species.

We know that this is happening in the human population because some groups are indeed reproducing more than others. First, birthrates are very low in developed countries and are continuing to fall. The population in Japan is currently shrinking. The populations of several Western European countries, such as Italy, France, and Austria, would be doing the same were it not for immigration. This means that the contribution of the Japanese and ethnic Central and Western Europeans to the future gene pool of the species is getting smaller and smaller.

Second, within a given country, whether developed or underdeveloped, some people reproduce more than others. This is not random. People with higher socioeconomic status have better access to education and more resources for birth control, both of which tend to be correlated with smaller family sizes. Many choose to forgo reproduction altogether. This ends up meaning that people with lower socioeconomic status tend to leave more offspring than do richer, more educated people. That could be considered a form of evolution too.

Besides economics, things like religion, education level, career advancement, family background, and even political beliefs all affect reproductive rates. In the West, these many factors influencing reproduction do not break down evenly across the various racial and ethnic groups because of the long history of racial oppression and ongoing social and political structures that reinforce inequality. This means that, in North America and Western Europe, people of African and Latino origin tend to have more children than nonimmigrant Caucasians. But even this trend is not uniform, and there are strong regional differences, making it all but impossible to predict where these evolutionary pressures are directing the species as a whole. Even the trends themselves are fluid.

In Asia, too, there are wide differences between reproductive patterns in different regions. Large families are completely unheard of in China, Japan, India, and most of Southeast Asia, while countries like Pakistan, Iran, and Afghanistan have sky-high birthrates.

Over time, these differences in birthrates will change the ethnic breakdown of our species. They also prove that the reproductive success of these various human ethnic groups is nonrandom—a precondition for evolution. It's true that differential *survival* is not a major phenomenon, at least not in the developed West, but differential *reproduction* certainly is. It doesn't matter that the differences are due to conscious reproductive choices—it's still unequal reproductive success. That's evolution.

Where is all of this leading? It's hard to say, but it's worth pointing out that racial and ethnic groups that had once been mostly isolated from one another are now in contact as never before, and intermarriage is happening at increasing rates. This could lead to the merging of the human species back into one interbreeding population—something that has not occurred, in all likelihood, since our species first originated in a small corner of Africa a couple of hundred thousand years ago.

Beyond that possibility, one thing *can* be said with absolute certainty: The only constant in life is change. To see how true this is, you need only look to the stars.

## Are We *Really* the Best Nature Can Do?

Enrico Fermi is one of the most important figures in modern nuclear physics. Among the many programs he was involved in was the Manhattan Project, in which he helped establish the conditions for a sustained nuclear reaction, a key component in the atom bomb. During a visit to Los Alamos, where the first bomb had been built less than ten years before, Fermi joined a casual conversation with Edward Teller and other scientists around a lunch table. This was at the height of the 1950s space race, and they were discussing the physical and technical barriers to traveling at near light speed. Most of the scientists eventually agreed that such rapid transit would one day be invented, and the conversation turned to guesses about when, not if, humans would achieve these great speeds. Most of those around the lunch table assumed that it would be a matter of decades, not centuries.

Suddenly, Fermi did some quick calculations on a napkin demonstrating that the galaxy had millions of planets similar to Earth. If interstellar travel was theoretically possible, then—"Where is everybody?" he suddenly blurted out.

The shocking realization that Fermi had while chatting over lunch that day was that the universe was eerily absent of non-natural radio signals. He and other scientists had been analyzing electromagnetic waves throughout the cosmos for many years. They had detected signals from very far away—millions and billions of light years away. But they had heard only the regular, repetitive signals from stars and other celestial bodies. They'd never heard anything that could possibly be a form of communication, as far as they could tell.

That realization was more than sixty years ago, and we still haven't heard anything but the background buzz of stars, planets, quasars, and nebulae, nor have we been visited by alien life (that we know of). Which leaves us with the uncomfortable question: If we do turn out to be the only intelligent life in the universe, what does that say about life in the first place — and what does it say about *us?*

As Fermi knew, the universe is billions of years old and contains billions of galaxies. Even our own Milky Way, a run-of-the-mill spiral galaxy, contains hundreds of millions of stars, each one of which could be orbited by a planet harboring intelligent life. Furthermore, from what we can tell from the fossil record, life on Earth started almost as soon as conditions were favorable. There was very little delay after the Earth cooled before life began buzzing along, well on its way toward evolving into complex organisms. This argues that life not only can evolve but *will* evolve on a lifeless planet when the temperature and chemical composition are right.

The vastness of the universe inspired Dr. Frank Drake to develop a mathematical formula, now known as the Drake equation, that attempts to estimate how many civilizations exist in the universe. There are many variables in the Drake equation, including the number of galaxies in the universe, the number of stars per galaxy, the rate of new star formation, the percentage of stars that have planets, the fraction of those planets that exist in the habitable zone (which allows liquid water), the odds that life will begin, the chances that life will evolve to the point of having intelligent beings capable of transmitting signals into space, and so on. None of these are perfectly knowable variables but all can be estimated using current knowledge and the laws of probability. While there are huge disagreements about the utility of the Drake equation, some current estimates predict that the universe harbors around seventy-five million civilizations. The estimates constantly change, of course, as our knowledge of the universe improves.

Even before the Drake equation was articulated, with so many billions of stars and planets, Fermi reasoned that the universe should be teeming with life. Moreover, alien civilizations could be very far ahead of us in terms of technological development. Most sci-fi movies imagine aliens that are just a couple hundred years beyond where we are now, but the universe is nearly fourteen billion years old, and stars and planets have existed for most of that time. Our solar system is relatively young at 4.6 billion years of age. There could be civilizations with *billions* of years on us in terms of their technology. They might be able to travel enormous distances the way we travel through cities.

Enrico Fermi's question became known as the Fermi paradox, summed up as "In a universe as old and vast as ours, why have we not yet heard from alien life?" There are many possibilities to this as yet unanswered question.

One possible explanation is that alien civilizations take care to hide their presence from us. An extreme expression of this notion is the planetarium hypothesis, which states that some sort of protective sphere has been built around us to filter out the noises from extraterrestrial civilizations but allow the background cosmic signals through.

Even if advanced alien civilizations have the ability (and desire) to keep us from hearing them, they would certainly be able to hear us. After all, we have been transmitting radio waves into space continuously since the 1930s. Traveling at the speed of light in all directions, our transmissions exit the solar system within a couple of hours and have been reaching other stars and their planets for decades. There are at least nine stars within ten light years of Earth and at least one hundred stars within twenty-five light years. Though our signals would be very weak by the time they reached that far, we would expect that an advanced civilization would also be advanced in its ability to monitor the signals coming from the surrounding stars and galaxies. They would know that we exist as well as quite a bit about us. (I wonder if that's why no one's come.)

Another explanation is that our assumptions are wrong and life is exceedingly rare in the universe. Maybe the rapid germination of life on Earth was an incredibly unlikely fluke and the other rare places that have been so lucky are so far away that radio signals have not had time to travel between them and us. Nevertheless, just in our local neighborhood of the Milky Way, we know that there are hundreds of thousands of planets in the sweet-spot range of temperatures necessary to sustain the kind of chemistry we see on Earth. Planets with the chemical composition of Earth and near the same temperature range are pretty common in the universe. While we don't have nearly enough information to determine much about what those planets are like, there is no reason to think that Earth was special in any way at the time that life got started.

Perhaps the most boring possible explanation is that every science-fiction book and film is wrong and our current barriers to interstellar travel are ultimately insurmountable. Stars are very far from one another, and at this time, we know of no way to exceed the speed of light—or even get anywhere near it. In fact, the conversation in which Fermi raised his question was actually an argument about the odds of humans having vehicles that could approach light speed in ten years' time. Fermi guessed 10 percent. That was more than sixty-five years ago, and we're no closer now than we were then to traveling at speeds anywhere close to light speed. If there simply *aren't* any solutions and standard jet propulsion is the best we're ever going to do, the many civilizations throughout the universe are destined to remain isolated from one another forever. We will stare at the stars, bored and lonely, while other beings stare back, but we'll never actually meet each other.

But still, why aren't we at least hearing their signals?

There is another explanation, an even darker one, that many scientists, myself included, are starting to worry about. It may be that life is relatively commonplace in the universe, but it appears—and disappears—over unfathomably immense timescales during periods that rarely if ever overlap. In other words, advanced alien civilizations are not out there waiting to be found because they no longer exist. And in all likelihood, the fate that befell them will also befall us: developmental implosion.

Think about it: Human beings are on a collision course with our own industrialization. We are consuming nonrenewable (or very slowly renewable) resources at an unsustainable pace. Coal, oil, and gas are finite resources. Even if there is a lot left, there is not an *infinite* amount left. We are converting rainforests, which produce the majority of our breathable oxygen and consume the majority of the carbon dioxide, into land for farming or housing. Our population is growing so fast that our ability to provide food for each person will be in serious doubt within a generation, despite all of our scorched-earth efforts to extract more and more sustenance from the planet. Meanwhile, climate change is threatening major coastline developments, some ocean ecosystems are in all-out collapse, and biodiversity throughout the globe is plummeting. We are in the midst of a mass extinction caused almost exclusively by our own actions. Who knows how bad things will get before we bottom out?

That's not even the worst of it. Weapons of mass destruction have raised the specter of mutually assured destruction, which was actually a delicate deterrent for a time but may not hold for long. Radical messianic and apocalyptic ideologues may be impervious to deterrence, and it seems inevitable that they will one day get their hands on the ultimate weapon. What will restrain them from using it? In addition, when world resources become scarce, strife will abound. Strife brings forth the worst in us, and economic and cold wars culminating in hot wars seem almost certain—with the stakes far higher than ever before.

Add to these dangers the very good chance that a pandemic could strike at any point. Humans now exist in such density that infectious diseases spread like wildfire. When we add to this the ease of global travel, a doomsday scenario is not hard to imagine.

All of these factors compound the others, increasing the danger that one or another of these tragedies will someday occur. Lack of farmable land raises food prices. Strain on energy resources raises *all* prices. High prices cause strife and unrest, which tends to favor the rise of dictators. Global warming will place the most pressure on the least developed regions, exacerbating their problems. Continued invasion of the rainforest will tap previously dormant viruses and provide them with a new and densely populated host. Bringing all of this together produces a grim picture. Are we on a clear path to our own demise?

There are literally thousands of ways to imagine how our species could suffer tremendous setbacks in the coming century, but at this point, extinction of *Homo sapiens* seems very unlikely. Given that humans live basically everywhere on the planet, there will always be people with the forethought, tenacity, and luck to get through whatever crises may come. Sure, without a complete redirection of our species' current trajectory, major economic and political collapse may be likely. But I have little doubt that some humans would survive an apocalyptic scenario and the species would go on, even if the destructive implosion causes mass death and suffering as well as huge setbacks in technology and development.

The risks we now face as a species—dangers that are completely attributable to our own ambition—might very well be the normal way of things in the universe. If life has emerged on another planet, we can only assume that natural selection shapes that life more or less the same way it does ours. This is because natural selection is an extension of simple logic: Those who survive and reproduce well will leave more offspring than those who do not. It's hard to imagine life working any other way on another planet, despite how different everything (and everyone) might appear on the surface. However, never have we seen — and never, alas, could we predict — the evolution of disciplined self-control, long-range foresight, rampant selflessness, generous self-sacrifice, or even something as simple as willpower. Evolution has never shown an ability to plan ahead more than a generation or two.

Evolution has made us entirely selfish. Of course, as a social species, we have an expanded sense of self that includes children, siblings, parents, and anyone else we are closely affiliated with. We make sacrifices for our children because we see them as part of "us." But with this expanded sense of self comes limits. Our siblings and even our friends might be "us," but certain strangers aren't. Maybe we can expand further and say that people of one's race, religion, or nationality constitutes "us," but that still leaves a "them." In the same way that humans evolved to feel parental love, we evolved to hate or fear those who are not "us." This is true in all social mammals, so we have every reason to believe that life on another planet would follow this same logic.

It may be that we have never seen, heard, or been contacted by aliens because their civilizations simply collapsed under the weight of their own selfishness, technological advancement, and a host of other exacerbating factors before they gained the ability to leave their solar system. We ourselves are tantalizingly close to unlocking the secrets of space travel, harnessing endless energy from the sun, and keeping our bodies healthy indefinitely, but we also may be just as close to catastrophic collapse. Perhaps the same scenario repeats itself throughout the history of the universe, endless boom-and-bust cycles where a civilization almost takes the crucial next steps before blowing itself back to the agrarian days (if it's lucky) and starting all over again.

Our impending collapse may be unavoidable, given our evolutionary design. Our desires, instincts, and drives are the product of natural selection, which does not make long-term plans. Chaos, death, and destruction may be the true natural state of the universe and of all species, including ours. To borrow a quote from the legendary science-fiction writer Arthur C. Clarke, "Two possibilities exist: Either we are alone in the Universe or we are not. Both are equally terrifying."

## Immortality Within Our Reach?

Death is a fact of life for every living thing, and humans are no exception. Nevertheless, humanity has been obsessed with death and how to prevent it—or at least delay it—since the dawn of time. The oldest recorded story in the world, *The Epic of Gilgamesh,* is about the hero's quest for everlasting life. In the West, the legends of the philosopher's stone, the Fountain of Youth, and the Holy Grail center on the secret to immortality. In the East, foundational stories in Hinduism (the Amrita elixir), Chinese medicine (Ling Zhi, the supernatural mushroom), Zoroastrianism (the sacred drink Soma), and many others center on magic that promises everlasting life. Even the Greek word *nektar,* from the legend of the nectar of the gods, translates literally to overcoming (*tar*) death (*nek*).

If we can't stave off death, we can at least deny its obliterating effects. Most mythologies and religions center on the afterlife, an abstraction drawn from the very human refusal to believe that this life is all there is and we'll never see our lost loved ones again. But ironically, this widely shared belief in an afterlife has done little to stem the pursuit of everlasting life. (Is it not odd that Juan Ponce de Léon's desire to find the Fountain of Youth was in no way tempered by his devout Catholicism, a belief that promised him everlasting life already?)

Human technology—medicine and alchemy back then and, in modern times, engineering and computing—has been heavily focused on prolonging life. Immortality has always been the biggest prize, and countless prophets, kings, heroes, deities, and adventurers have taken enormous risks in pursuit of it. Today, for the first time, perpetual life may actually be within reach.

Science has been working hard to reveal the underlying mechanism of aging. As with all things in biology, the process is far more complex than we ever thought. Early research into aging revealed the discouraging truth that aging is caused by the accumulation of random damage to DNA and proteins. I call this discouraging because random damage is very hard to prevent. Modern medicine's ability to repair damaged tissue is laughable compared with the ability of the body to heal itself. If our own bodies cannot figure out how to stop the cumulative onslaught of molecular damage, what hope do our brains possibly have? The damage is not on the microscale but on the *nano*scale, and our blunt instruments can barely see it, let alone repair it.

Nevertheless, new, wholly different strategies to prolong life are beginning to emerge. For one thing, science has wisely given up on the idea that cellular damage is something that physicians can repair. Instead, efforts have largely been focused on understanding how stem cells work and determining whether they can be harnessed. Stem cells are the body's built-in renewal system for tissues. Stem cells, while few in number, are dispersed strategically throughout most organs and usually lie dormant until called upon. When specialized cells are lost to injury, illness, or mutation, stem cells jump into action and proliferate, producing replacement cells that can then differentiate into specialized cells and begin to function.

Scientists are finding stem cells in every tissue they examine, and it turns out that the body is more capable of self-renewal than previously thought. It was once considered a canonical truth that every human is born with all the neurons that he or she will ever have and that the gradual loss of neurons in aging adults is inevitable and irreversible. It turns out that the brain has neuronal stem cells that can replace lost or damaged neurons in specific instances. While the information stored by a lost neuron is probably lost forever, the brain does appear able to grow new neurons.

Stem cells are thus one avenue through which biomedical scientists are attempting to prolong human life indefinitely. If they can figure out how to enhance human stem cells so that they don't lose the race against cellular damage, we'd have a real chance at living much longer.

But other, more science-fiction-inspired efforts to prolong life are also under way. Technologies around tissue and organ transplants are developing so rapidly that doctors will soon attempt the transplant of a human head. Actually, this frames it backward. Because an individual's personality, memories, and consciousness are housed wholly in the brain, this procedure should really be considered a body transplant. If these transplants become successful and efforts to keep brain tissue refreshed and functioning are successful, a person could live indefinitely by simply transplanting her head from body to body. (Let's not stop to worry about where the bodies would come from.)

An even more futuristic but perhaps more realistic possibility is the continuing development of xenobiotic and synthetic bionic implants. Beginning with horsehair sutures to close wounds in ancient times and continuing with hooks and peg legs to replace lost appendages in the Middle Ages, humans have long sought to overcome biological limitations with synthetic alternatives. More recently, physicians have advanced from using pig parts to replace failing heart valves to using artificial valves that are sure to outlast their recipients. In fact, scientists have now developed a whole artificial heart that can completely replace its biological counterpart.

While the current limitations of artificial hearts means that the recipients must await the more lasting solution of a transplant, people have gone for years with something called a left-ventricular assist device, which almost completely takes over the pumping function of the heart. Just a few decades ago, who would have thought that someone could live indefinitely and nearly symptom-free after his own heart had almost failed? That's exactly what former U.S. vice president Dick Cheney did until he finally received a heart transplant.

Even our current arsenal of bionic implants reads like the science fiction I grew up reading in the 1980s. Cochlear implantation is now routine, as are arterial stents, artificial hips and knees, and glucose monitors paired with insulin pumps. Already on the horizon are artificial eyes to send visual information directly to the brain, like Geordi La Forge on *Star Trek: The Next Generation*. The big breakthrough will likely come as we pair our understanding of tissue renewal with the capabilities of nanotechnology. We have nearly all the tools and knowledge necessary to engineer tiny repair robots to scavenge organs for aging cells and recruit fresh stem cells to replace them. It's only a matter of time.

Eventually, we may not even need to go to all of that trouble. A new technology called CRISPR/Cas9 has revolutionized science's ability to safely edit the DNA of living cells. Until recently, the promise of gene therapy was limited by practical difficulties. It seemed impossible and had proved unsafe in even modest attempts. However, CRISPR has changed all of that, and the means to slice and dice our genomes appears tantalizingly close. Biomedical scientists in all fields are racing to see if—or, rather, how—CRISPR can be used to cure disease, repair damage, and renew tissues.

Genetic testing and counseling have already affected human evolution in this regard. Many people who have a history of certain genetic diseases in their families or ethnic backgrounds elect to get genetic counseling. Couples who discover that both are carriers of a serious genetic illness can choose to go their separate ways, eschew having biological children, or employ amniocentesis to detect whether a fetus will have the dreaded malady. The effect of these efforts is that the prevalence of these diseases in the population is decreasing. This phenomenon will likely be further enhanced through CRISPR. Couples wishing to have a child may one day be able to have their eggs and sperm not just analyzed but repaired prior to fertilization. CRISPR could slice out the disease-causing version of the gene, replace it with the healthy version, and voilà! The technology to do this already exists and will no doubt be tested in fertility clinics soon.

Even more incredibly, in addition to fixing genetic diseases, CRISPR could easily be used in sperm and eggs to alter the genetics of the planned child to extend his or her lifespan even further. As we come to understand the genetic control of aging, scientists may one day be able to make tweaks to the genes of future generations such that they don't grow old in the first place.

Of course, as I noted earlier, the real prize is the search for immortality. As the full picture of cellular aging and tissue renewal comes into focus, we may be able to deploy CRISPR-armed nanobots to fix damaged cells before they begin to show their age. This is not wild speculation. The first steps toward this approach are already being envisioned in animal models. Yes, the initial attempts will be modest, but if they are successful, this genie will never go back in the bottle.

All the technology I've discussed here is nearly at hand, and its arrival in your doctor's office may be only a few decades away. Certainly, medical technology for prolonging life is developing rapidly even by conventional standards, and for people who manage to stay alive until the deployment of these new measures, doctors may be able to stop, or at least slow, the hands of time. As the technology continues to develop, as it surely will, doctors may even be able to reverse (not just pause) the effects of aging, and individuals will get to live as twenty-somethings forever. This notion is driving many of those near the middle of their lives, including myself, to get in shape and try to "live long enough to live forever," the prescient subtitle of a book published in 2004.

Where we will fit all these newly immortal people is another question — but given our species' tendency to kill one another in large numbers, this problem may solve itself when resources become scarce. Another possibility is the colonization of other planets and moons in our solar system or nearby ones. While this may seem far-fetched because aerospace technology has not developed as rapidly as biomedical technology has, we may be approaching a watershed moment on that front as well.

The upshot: Never underestimate science or our species' ability to overcome its own flaws. In fact, many anthropologists credit the development of humans' impressive ingenuity to the dramatic climate changes that occurred in Africa, Europe, and Central Asia over the past two million years. Biology alone could never have gotten humans through the ice age. We needed cleverness too. And today we are in desperate need of that crucial quality, perhaps more than ever before.

## Coda: Swords or Plowshares?

While no one knows for sure what the future holds for humanity, we can get an idea by looking at our past. We are a beautiful but imperfect species. What has defined our past will define our future. Because the past is filled with stories of struggle and misery giving way to triumph and prosperity, there is hope that the same will be true for our future. The struggle is clear: Our population growth, environmental destruction, and poor stewardship of natural resources threaten the prosperity that we have sought to create for ourselves.

What is the answer to that struggle? How can we turn the impending doom into triumphant peace? Simple. By using the same tools and processes that helped us overcome our previous challenges, the same means that brought us prosperity and abundance in the first place: science.

You might be thinking, *Maybe science itself is the problem. Maybe our reliance on science and technology is our ultimate flaw.* That's an understandable suspicion. But I don't think it's the reality.

It is true that scientific progress led to the development of the coal- and petroleum-based energy industry that is devastating the carbon balance in our atmosphere. But science has also provided the solution: solar, wind, water, and geothermal power. It is true that agricultural and textile technologies have led to massive deforestation and tremendous pollution from factory farms. But science has also cultivated the clean crops and synthetic alternatives that could one day phase out their polluting predecessors. The same commitment to scientific progress that conjured coal-powered steam engines has now developed a solar-powered airplane. While every piece of plastic ever made is now sitting in a landfill or on its way to doing so, chemists have created biodegradable plastics, and biologists have engineered bacteria that can eat plastic. Every problem that science has created, science can solve.

If that sounds overly optimistic, consider this: Green buildings are going up left and right, and we are increasingly meeting our demands for energy and materials in sustainable and environmentally friendly ways. Per square foot, the average American home runs on less than half the electricity annually than it did twenty-five years ago. Per gallon of gasoline, the average new car goes twice as many miles as it did thirty-five years ago. And for both homes and cars, solar and other carbon-neutral power are increasingly pushing down the demand for combustion-based energy. Several European nations have the goal of being carbon-neutral squarely within their sights, and these countries have nowhere near the harvestable sunshine that countries in the Global South have.

A better future is within our reach. The question is, will we be able to grasp it? Or, to put it a different way: Will our advanced intelligence prove to be our biggest asset or our biggest flaw?

We already have the science that can save our species from itself. We are waiting only for the will. And if we can't muster it in time to prevent a global collapse, we will have the ultimate proof of our poor design.

# Acknowledgments

This book has benefited from the hard work of so many whose names should rightly be on the cover. Marly Rusoff, you gave this project life. As with our previous book together, Tara VanTimmeren took the first pass on everything in this book. Only after it has benefited from her shaping and polishing do I ever dare send a manuscript to anyone else. From our first breakfast meeting, I knew that you were "the one" and I immediately trashed the list of agents to whom I had planned to pitch this. You took my scattered thoughts and helped me produce a coherent manuscript from them. Bruce Nichols and Alexander Littlefield, you both have been incredibly insightful editors whose contributions have improved this book tenfold. Thanks to all four of you talented editors for believing in this project and bringing the skill and professionalism needed to transform it from a nice idea to a finished book. Tracy Roe also deserves buckets of praise for her spectacular eleventh-hour contributions to this manuscript, which strengthened it immeasurably. This book really was a team effort and it was humbling to work with people of such intellect.

I must also acknowledge the work of the incredibly talented artist whose playful but illuminating drawings grace these pages. It was so gratifying to watch Don Ganley take my often vague and unhelpful guidelines and make wonderful illustrations from them. His art really brings the content of this book to life. I hope you take a minute to fully appreciate these drawings. Each one is the result of many hours and many revisions. It took Don like three hours to finish the shading on the upper lip of the skull in the figure on page 11. It's probably the best drawing he's ever done.

And to my students, my friends, and my family, thank you so much for letting me drone on and on about these topics over the years. I always strive for a writing style that is best described as a fun conversation with a friend; that is to say, I try to write as though I am talking with you. If you have ever indulged me in a conversation about any of these topics, you unwittingly helped me write this book. And for that, I cannot possibly thank you enough.

As with everything else I do, this book would not have been possible without the support of my family, whose patience was surely tested over the years that I, among the most flawed members of our species, worked on this manuscript. Oscar, Richard, Alicia, and, of course, Bruno, thank you for the encouragement. I love you.

# Notes

1. Pointless Bones and Other Anatomical Errors

2    *30 to 40 percent of the population:* Seang-Mei Saw et al., "Epidemiology of Myopia," *Epidemiologic Reviews* 18, no. 2 (1996): 175–87.

4    *migrating birds detect:* Thorsten Ritz, Salih Adem, and Klaus Schulten, "A Model for Photoreceptor-Based Magnetoreception in Birds," *Biophysical Journal* 78, no. 2 (2000): 707–18.

     *For a human to achieve:* Julie L. Schnapf and Denis A. Baylor, "How Photoreceptor Cells Respond to Light," *Scientific American* 256, no. 4 (1987): 40.

17   *how long the RLN must have been:* Mathew J. Wedel, "A Monument of Inefficiency: The Presumed Course of the Recurrent Laryngeal Nerve in Sauropod Dinosaurs," *Acta Palaeontologica Polonica* 57, no. 2 (2012): 251–56.

32   *Japanese fishermen caught a dolphin:* Seiji Ohsumi and Hidehiro Kato, "A Bottlenose Dolphin (*Tursiops truncatus*) with Fin-Shaped Hind Appendages," *Marine Mammal Science* 24, no. 3 (2008): 743–45.

2. Our Needy Diet

39   *the GULO gene suffered a mutation:* Morimitsu Nishikimi and Kunio Yagi, "Molecular Basis for the Deficiency in Humans of Gulonolactone Oxidase, a Key Enzyme for Ascorbic Acid Biosynthesis," *American Journal of Clinical Nutrition* 54, no. 6 (1991): 1203S–8S.

41   *Take fruit bats:* Jie Cui et al., "Progressive Pseudogenization: Vitamin C Syn-

thesis and Its Loss in Bats," *Molecular Biology and Evolution* 28, no. 2 (2011): 1025–31.

45  *in case you are wondering:* V. Herbert et al., "Are Colon Bacteria a Major Source of Cobalamin Analogues in Human Tissues?," *Transactions of the Association of American Physicians* 97 (1984): 161.

60  *flooded with diet books:* This section is adapted from a passage in chapter 8 of my first book, *Not So Different: Finding Human Nature in Animals* (New York: Columbia University Press, 2016).

62  *people who are active athletes:* Amy Luke et al., "Energy Expenditure Does Not Predict Weight Change in Either Nigerian or African American Women," *American Journal of Clinical Nutrition* 89, no. 1 (2009): 169–76.

## 3. Junk in the Genome

73  *Scientists estimate:* David Torrents et al., "A Genome-Wide Survey of Human Pseudogenes," *Genome Research* 13, no. 12 (2003): 2559–67.

74  *in an ancestor common to humans:* Tomas Ganz, "Defensins: Antimicrobial Peptides of Innate Immunity," *Nature Reviews Immunology* 3, no. 9 (2003): 710–20.

87  *Once upon a time:* Jan Ole Kriegs et al., "Evolutionary History of 7SL RNA-Derived SINEs in Supraprimates," *Trends in Genetics* 23, no. 4 (2007): 158–61.

## 4. *Homo sterilis*

105  *As of 2014:* All statistics from Central Intelligence Agency, *The World Factbook 2014–15* (Washington, DC: Government Printing Office, 2015).

108  *In chimps, the average spacing:* Biruté M. F. Galdikas and James W. Wood, "Birth Spacing Patterns in Humans and Apes," *American Journal of Physical Anthropology* 83, no. 2 (1990): 185–91.

117  *One study found an orca:* Lauren J. N. Brent et al., "Ecological Knowledge, Leadership, and the Evolution of Menopause in Killer Whales," *Current Biology* 25, no. 6 (2015): 746–50.

121  *If grandparental investment is so great:* This is somewhat disputed because there have been some reports of reproductive senescence in captive populations of primates and some other mammals; however, these isolated cases do not approach the universal and carefully timed nature of human menopause.

## 5. Why God Invented Doctors

135  *Indeed, many histories of sanitariums:* Norman Routh Phillips, "Goitre and the Psychoses," *British Journal of Psychiatry* 65, no. 271 (1919): 235–48.

140  *neutralize the invaders:* This is how vaccines work; when you're given an injection of a dead or damaged virus, your immune system learns how to fight it. If all goes well, the immune system is shaped so that the next time it sees that antigen — like when you are exposed to the actual virulent virus — it mounts a response that is hundreds of times faster and more vigorous than it would have been if it were seeing it for the first time.

144  *Prevalence of both food and respiratory allergies:* Susan Prescott and Katrina J. Allen, "Food Allergy: Riding the Second Wave of the Allergy Epidemic," *Pediatric Allergy and Immunology* 22, no. 2 (2011): 155–60.

### 6. A Species of Suckers

165  *rate the quality and relevance of each study:* Charles G. Lord, Lee Ross, and Mark R. Lepper, "Biased Assimilation and Attitude Polarization: The Effects of Prior Theories on Subsequently Considered Evidence," *Journal of Personality and Social Psychology* 37, no. 11 (1979): 2098.

*two political hot topics:* Charles S. Taber and Milton Lodge, "Motivated Skepticism in the Evaluation of Political Beliefs," *American Journal of Political Science* 50, no. 3 (2006): 755–69.

*Another manifestation of confirmation bias:* Bertram R. Forer, "The Fallacy of Personal Validation: A Classroom Demonstration of Gullibility," *Journal of Abnormal and Social Psychology* 44, no. 1 (1949): 118.

170  *This overremembering:* Steven M. Southwick et al., "Consistency of Memory for Combat-Related Traumatic Events in Veterans of Operation Desert Storm," *American Journal of Psychiatry* 154, no. 2 (1997): 173–77.

*Researchers led by my colleague:* Deryn Strange and Melanie K. T. Takarangi, "False Memories for Missing Aspects of Traumatic Events," *Acta Psychologica* 141, no. 3 (2012): 322–26.

177  *When you're in the middle of a losing streak:* The only time when this may not hold in a casino is in blackjack, when there are a finite and knowable number of face cards. A long stretch of non–face cards does indeed mean that the remainder of the shoe will be enriched with face cards. Of course, this could help the dealer as often as it does the gamblers, and there is no guarantee that the deficit will be made up before the cut card is reached and the last hand is called for that shoe. Nevertheless, a skilled card counter can give himself a slight advantage over the house that might reap financial benefits over a long day of card playing. However, casinos have ways of spotting card counters and will place the cut card very shallowly in the shoe to neutralize them. And if that doesn't work, the manager will show the card counter the door. The house always wins.

183  *Dr. Laurie Santos:* M. Keith Chen, Venkat Lakshminarayanan, and Laurie R. Santos, "How Basic Are Behavioral Biases? Evidence from Capuchin Monkey Trading Behavior," *Journal of Political Economy* 114, no. 3 (2006): 517–37.

### Epilogue: The Future of Humanity

214  *This notion is driving:* Ray Kurzweil and Terry Grossman, *Fantastic Voyage: Live Long Enough to Live Forever* (Emmaus, PA: Rodale, 2004).

# Index